IBERIAN AND LATIN AMERICAN STUDIES

Hermaphroditism, Medical Science and Sexual Identity in Spain, 1850–1960

Series Editors

Professor David George (Swansea University)
Professor Paul Garner (University of Leeds)

Editorial Board

David Frier (University of Leeds)
Lisa Shaw (University of Liverpool)
Gareth Walters (Swansea University)
Rob Stone (Swansea University)
David Gies (University of Virginia)
Catherine Davies (University of Nottingham)

IBERIAN AND LATIN AMERICAN STUDIES

Hermaphroditism, Medical Science and Sexual Identity in Spain, 1850–1960

Richard Cleminson and Francisco Vázquez García

UNIVERSITY OF WALES PRESS
CARDIFF
2009

© Richard Cleminson and Francisco Vázquez García, 2009

All rights reserved. No part of this book may be reproduced in any material form (including photocopying or storing it in any medium by electronic means and whether or not transiently or incidentally to some other use of this publication) without the written permission of the copyright owner except in accordance with the provisions of the Copyright, Designs and Patents Act 1988 or under the terms of a licence issued by the Copyright Licensing Agency Ltd, Saffron House, 6–10 Kirby Street, London, EC1N 8TS. Applications for the copyright owner's written permission to reproduce any part of this publication should be addressed to the University of Wales Press, 10 Columbus Walk, Brigantine Place, Cardiff, CF10 4UP.
www.uwp.co.uk

British Library Cataloguing-in-Publication Data
A catalogue record for this book is available from the British Library.

ISBN 978-0-7083-2204-8
e-ISBN 978-0-7083-2279-6

The rights of Richard Cleminson and Francisco Vázquez García to be identified as authors of this work has been asserted by them in accordance with sections 77, 78 and 79 of the Copyright, Designs and Patents Act 1988.

Typeset by Mark Heslington Ltd, Scarborough, North Yorkshire
Printed by CPI Antony Rowe, Chippenham, Wiltshire

Contents

Series Editors' Foreword	vii
Acknowledgements	ix
Chapter 1: Introduction: Male, Female or In-Between? Towards a History of the Science of 'Hermaphroditism' in Spain, 1850–1960	1
Chapter 2: From Sex as Social Status to Biological Sex	29
Chapter 3: Between Diagnoses: Hermaphroditism, Hypospadias and Pseudo-hermaphroditism, 1870–1905	78
Chapter 4: Gonads, Hormones and Marañón's Theory of Intersexuality, 1905–1930	122
Chapter 5: From True Sex to Sex as Simulacrum	179
Chapter 6: Conclusions	224
Bibliography	241
Index	265

Series Editors' Foreword

Over recent decades the traditional 'languages and literatures' model in Spanish departments in universities in the United Kingdom has been superceded by a contextual, interdisciplinary and 'area studies' approach to the study of the culture, history, society and politics of the Hispanic and Lusophone worlds – categories which extend far beyond the confines of the Iberian Peninsula, not only in Latin America but also to Spanish-speaking and Lusophone Africa.

In response to these dynamic trends in research priorities and curriculum development, this series is designed to present both disciplinary and interdisciplinary research within the general field of Iberian and Latin American Studies, particularly studies which explore all aspects of **Cultural Production** (inter alia literature, film, music, dance, sport) in Spanish, Portuguese, Basque, Catalan, Galician and indigenous languages of Latin America. The series also aims to publish research on the **History and Politics** of the Hispanic and Lusophone worlds, at the level of both the region and the nation-state, as well as on **Cultural Studies** which explore the shifting terrains of gender, sexual, racial and postcolonial identities in those same regions.

Acknowledgements

Richard Cleminson would like to acknowledge the intellectual stimulus freely given by Rosa María Medina Doménech, with whom the ideas that became this book were originally discussed, the searching questions posed at the Hispanic Studies seminar at the University of Cambridge, the interest with which some aspects of this project were received at a conference on the body and sexuality at the University of Exeter and the various conversations with Lena Eckert, Jennifer Jordan, Lesley Hall, Chris Perriam, Lola Sánchez and Alison Sinclair. He would like to thank the staff of many libraries, including those working in Document Supply at the University of Leeds, at the University of Granada and the University of Oviedo, the staff at the Covadonga hospital library, Oviedo, the Wellcome Library, London, especially Venita Paul, and the library of the Hospital San Juan de Dios, Granada, especially Gustavo Zenner. In particular, like Francisco Vázquez, he would like to thank Ana Remón, the director of the Faculty of Medicine, University of Cadiz library. Part of the funding for completing this research came from the Arts and Humanities Research Council under its Research Leave scheme. Without this funding, the completion of this project would have taken much longer, if achieved at all. Some of the initial research was undertaken while in receipt of a Wellcome Trust History of Medicine Award for a previous project on male homosexuality in Spain. Final thanks go to Fredy Vélez for unfailing support, wine and dinners and company on the way.

Francisco Vázquez would like to acknowledge the assistance of many individuals who have sought out and made available rich materials in order to complete this book. These include: Chema Fraile, María Jesús Ruiz, Arturo Morgado and José María López Cepero. I thank the historians María José de la Pascua, Mónica Bolufer and Andrés Moreno Mengíbar for what their writings have taught me and for their helpful comments. Particular thanks go to the director of the library of the Faculty of Medicine at the University of Cadiz, Ana

Remón. My colleagues in philosophy have provided an ideal place in which to enjoy an excellent and pleasant environment to work. I would also like to thank the Department of Spanish, Portuguese and Latin American Studies at Leeds University, the research group 'Ubi Sunt', Rafael Vélez and Francisco Ortega Guerrero, who invited me to talk about some of the ideas contained in this book, as well as the students for the Master's degree in Gender and Citizenship at the University of Cadiz for their suggestions and interest. Finally, thanks go to Oliva and Curro for their generosity and constant support.

Both authors would like to thank Sarah Lewis, Elin Nesta Lewis and Siân Chapman of the University of Wales Press for their work, encouragement and patience over the course of this project.

Note: The numerous minor errors of Spanish in the original texts have been left uncorrected. All translations, unless otherwise stated, are our own.

Chapter 1
Introduction: Male, Female or In-Between? Towards a History of the Science of 'Hermaphroditism' in Spain, 1850–1960

Methodological and theoretical considerations

At least since Ovid's account in book 4 of his *Metamorphoses*, in which the gods Mercury and Venus, embodiments of ideal manhood and womanhood respectively, had a son named Hermaphroditus, hermaphroditism, or the apparent mixing of the sexes, has been a subject of fascination for the West.[1] Hermaphroditus was, like his parents Hermes and Aphrodite in their Greek incarnations, a perfect example of humankind. His exemplary male physique, however, was to be transformed during his encounter with the nymph Salmacis who requested that the gods unite her forever to the boy. Hermaphroditus' encounter with Salmacis resulted in Hermaphrodite, a body with double sex or 'in-between', 'weakened' status, an 'androgyne son'.[2] After his transformation, Hermaphrodite requested that any future male who passed by the nymph's pool where he was changed should likewise emerge with his 'manhood diminished'.[3] Such a story fired the western literary and medical imagination both for its attractiveness – the harmonious combination of two 'opposites' in one body[4] – and for its marvellous but unsettling nature – the mixing of the two sexes which might signify great good or evil to come, a dual perception that has continued to inform historical accounts of hermaphrodites and the related fields of androgyny, intersexuality and transsexualism to this day.[5]

Examples of 'hermaphrodites' have recurred in different countries over time. Principally, they have been the focus of attention of the religious authorities and the medical and legal professions, which viewed the hermaphrodite as a disruption of what might now be termed accepted gendered and sexual norms. Examples of this disruption and the ways in which such figures were dealt with by the authorities include the case of Marie/Germain Garnier discussed by the French royal surgeon Ambroise Paré in 1573,[6] the case of Helena de Céspedes heard by the Toledo Offices of the Inquisition in 1587,[7] and the nineteenth-century French Herculine Barbin, all discussed below.[8]

What was at stake for these medical and/or religious authorities was the construction of an account that satisfactorily aligned body, gender and sex into what was for them a harmonious whole, eliminating or explaining the 'abnormal'.[9] This was not, however, a static process. Changing medical diagnoses and social expectations with regard to the sexes meant a constant rewriting of the interrelationship between the body, conceptions of the sex of the individual and of what we now term gender and sexual preference as a means of situating bodies that did not fall clearly into the male/female divide. These changing conceptualizations beg a number of questions: what were the criteria according to which the identification of the sex of the person was made? Why was it necessary to determine the 'true' sex of the individual whose masculinity or femininity was in doubt? To what degree did the changing contested social positions of the sexes, in the context of new marriage legislation, for example, affect doctors' discussions of the 'real sex' of the individual? How did the concerns of new nineteenth- and twentieth-century liberal philosophies and regimes influence medical perceptions of sex?[10]

This book will focus primarily on medico-legal discourse on the hermaphrodite in Spain and will argue that the 'science of hermaphroditism', emerging from the mid-nineteenth century onwards, was an attempt by medical and legal authorities to manage a cluster of 'deviant' representations and acts around body, sex and gender in social and medical circumstances that changed over time.[11] This science-in-construction must be considered as part of other emerging sexual sciences of the period, particularly the inquiries into homosexuality and other forms of gendered and sexual 'deviance'.[12] As is the case for homosexuality, the science of hermaphroditism underwent numerous changes throughout the nineteenth and twentieth centuries. This book sets out to chart and explain them.

Historians generally agree that changes in the conceptualization of the hermaphrodite have passed through the following broad stages from the medieval period: a contested but generally hegemonic acceptance of the hermaphrodite as a kind of intermediary between male and female emerging out of concurrent but conflicting accounts stemming from the thought of Hippocrates, Aristotle and Galen, amongst others, on the generation of humans and animals up to the early sixteenth century; a period of 'disenchantment' when the hermaphrodite's marvellous nature and actual possibility were steadily rejected; and his/her medicalization and ultimate practical impossibility by the end of the nineteenth century. By the beginning of the nineteenth century doctors faced with cases of ambiguous sex argued that hermaphrodites did not in fact exist among human beings save in extremely rare cases and that individuals of doubtful sex were in reality *apparent*, not *real* hermaphrodites.

Accompanying this process of the elimination of the possibility of the real physical hermaphrodite was the concomitant and parallel emergence of a different and new sexual figure. After his/her many transformations over time, from the mid-nineteenth century the hermaphrodite took two principal routes: first, the anatomical hermaphrodite, who was gradually given less and less discursive space, and, secondly, the so-called 'psychic hermaphrodite', the category that was to become the 'homosexual'.[13] The two were never far apart, however, and hermaphroditism remained as a 'ghost category' in the construction of sexual inversion and homosexuality, sometimes coming more to the fore and on occasion shrinking into the background.[14]

Such an outline of the genealogy of the hermaphrodite does entail certain problems of interpretation. Rather than trace an apparently continuous and coherent line between documented historical cases of supposed hermaphroditism from the sixteenth century onwards and those persons deemed 'intersexuals' in the period studied in this book, we need to be aware that 'hermaphrodites' in the past may have not displayed the same characteristics as those that we discuss here. The very diagnostics of the category and the category itself have changed. The term 'intersexual' did not exist until the early twentieth century and 'transsexualism' as a 'condition' where sex change was sought was first elaborated in 1949; neither meant the same as the 'hermaphrodite' of yesteryear.[15] On the one hand, therefore, it is necessary to avoid the culturally conditioned assumptions adopted by some writers who study the hermaphrodite, the androgyne or the intersexual as an unproblematic continual presence in history or in societies outside of the west, imbuing him or her with particular qualities and with connections to a supposed transnational hermaphrodite identity.[16] On the other hand, we are conscious that, precisely because similar terminology has been used to designate people of an intersexual condition, we cannot erase any kind of continuity between seemingly disparate examples. The question is therefore posed: how do we write historically about 'hermaphrodites'?

In what follows we propose a four-part discussion as a prelude to our analysis of the hermaphrodite in Spain in the period 1850–1960. First, we review a number of historians' accounts of hermaphroditism in history. In this section, the legacy of material on the subject from the medieval period, the relative importance of Hippocratic and Aristotelian perspectives on hermaphroditism, Foucault's writing on hermaphroditism and Thomas Laqueur's work on changes in western understandings of the differences between male and female are key subjects. Secondly, we discuss some theoretical debates within feminist studies with respect to the relationship between the body, gender and sex. This analysis complements particularly Foucault's analysis of hermaphroditism by introducing an account of notions of biological 'sex' from a feminist perspective that deconstructs concepts of nature and nurture, the fixity or otherwise of sex and their relationship to gender and the body. Thirdly, we outline a possible framework for the consideration of hermaphroditism in Spain by drawing on the work of Alice Dreger and Nelly Oudshoorn with respect to the stages of medical development of hermaphroditism and the establishment of endocrinology and hormonal analysis.[17] Finally, we dedicate a section to the state of current historical debates in Spain on the figure of the hermaphrodite and with special reference to our period, 1850–1960. The first three sections are meant as debates; rather than adopting one approach we will consider our subject from a variety of standpoints as we progress.

Hermaphroditism and intersexuality: historical perspectives

Western thought on the subject of hermaphroditism stems largely from Graeco-Roman treatment of hermaphroditism and the natural scientific work of Hippocrates, Aristotle and Galen on the generation of human beings and the fixing of sex in medieval medical thought.[18] While the influence of these sources on medieval and later medical thought is extensive it is clear that many basic concepts from this time survived into the modern period; it is not a case of an abrupt change being effected in studies of hermaphroditism come the nineteenth century but rather one of uneven development with regional idiosyncrasies.[19]

In first-century BCE Greece there were two principal competing theories on the generation of human beings, each providing a more medicalized rather than a mythological interpretation of the ancient Graeco-Roman myths such as that of Hermaphroditus and Salmacis or the interpretation of love and sex advanced by Plato.[20] The first of these theories was articulated by Hippocrates (c.460–375 BCE), who believed that human beings were not divided into two sexes, but that there existed one sex, the male sex, with derivations of varying perfection from this model. The most perfect representation was the virile male and the most imperfect the completely feminine female. In between these poles there existed, according to the physician, a whole range of intermediate figures and sexual transmutations, including menstruating males, bearded women, viragos, effeminates and different types of hermaphrodite. Hippocrates believed that the sex of the foetus was determined by two opposites: the maternal and paternal principles that each produced their kind of seed. In accordance with the position of the foetus in the womb, which was divided into different 'chambers', and with the relative domination of male or female seed, the new-born child would either be male or female or somewhere in between. If the foetus was located in the centre of the womb and received equal doses of male and female principles it would be truly hermaphroditic.[21]

Thomas Laqueur has argued that this Hippocratic model dominated western notions of the generation of the sexes and, by connection, of hermaphrodites until the eighteenth century. This 'one-sex' model, whereby the fully virile male represented perfection and all other derivations were deemed less perfect, was one where society would still speak of men and women but instead of major differences between them in terms of biology, what counted essentially were differences in terms of the social role of men and women. This 'one-sex' model would give way to a 'two-sex' model in the eighteenth century, in which males and females were seen as different sexes in themselves as part of an Enlightenment project which deemed females equal but different in social and biological terms and which underpinned the separation of the public and private spheres as part of western democracy.[22] This acceptance of fundamental differences between the sexes was in part a response to medical developments but was also due to broad changes in the social world resulting in clearer roles for gendered behaviour. Women came to be associated fundamentally with their 'sexual nature' and would be dedicated to childbearing and nurturing while the male would transcend his sexual nature,[23] although 'sexually deviant' males would be excepted from this model of self-control.

Other historians, however, have questioned Laqueur's account. Joan Cadden has argued not for the dominance of one theory of generation (Hippocrates') over another in medieval times (eleventh–fourteenth centuries) but in favour of a wide variety of views on the generation of the sexes.[24] Daston and Park note that the tradition of theories of generation was 'more complicated and internally diverse' than Laqueur's depiction. They point to our second theory on the generation of animals and sex: that of Aristotle.[25] Aristotle (384–322 BCE), author of *De Generatione Animalium*, argued in favour of a dichotomous nature of the sexes, not a continuum or a scale from full male perfection to female imperfection.[26] In the same way as other Aristotelian polarities such as active/passive, perfect/imperfect and matter/form were held to exist, Aristotle did not admit intermediate or transitory stages between the sexes. For this reason, Aristotle believed that real hermaphrodites did not in fact exist. Declaring in addition that female seed did not exist, in contrast to Hippocrates' understanding, any individual not easily classified as male or female would result when the amount of matter contributed by the mother was *excessive* for the growth of just one fully formed foetus. 'Hermaphroditism', then, was no more than a question of 'extra sex (genital) parts added on to their single "true" sexes'.[27] In this model, hermaphroditism was therefore like 'extra toes or nipples, in that it represented an overabundance of generative material',[28] a kind of deformation but not an intermediate sex.

Such a view was largely accepted, despite a number of modifications, by the physician Galen (131–201 CE). Galen relied on Aristotle's writings in order to establish his own theories on the generation of human beings. The physician agreed with Hippocratic sources that women were moister than men but accepted, following

Aristotle, that women were cooler and men warmer. Even though Galen disagreed with Aristotle, however, on the question of seed (Galen followed Hippocrates on the 'two-seed' model rather than Aristotle's acceptance of male seed only), he did adopt the polarity of left and right shared by both Aristotle and Hippocrates and he employed the traditional association of right with male and left with female. Further, in an additional twist and in contrast to Aristotle, he did not use this device to emphasize any differences between the sexes but argued, this time following Aristotle, that females were less perfect than males because they were less warm. Aristotle used the female lack of heat as an explanation of female inability to produce semen, thus reasserting the contrary nature of the sexes and underscoring the value of social and sexual hierarchy. Galen, on the other hand, took the Hippocratic position, arguing that females did produce semen but that it was less plentiful and was cooler and moister than males'. Finally, male and female parts were anatomically equivalent for Galen but viewed through an internal/external prism, that is, what the male possessed externally, the female held internally.[29]

How did these various competing or complementary theories affect notions of hermaphroditism in the West in the ensuing centuries? Over a period of a thousand years, 'these two contrasting, ancient accounts of hermaphroditism were transmitted to medieval and early modern medical theorists in a number of stages, both directly – as various of the key texts were successively translated from Greek into Latin – and indirectly, through the intermediary of Arabic writers such as Avicenna', Daston and Park argue.[30] The result of this eclectic mix was a complex and uneasy juxtaposition or fusion of theoretical positions with the Hippocratic model dominating until the early middle ages. With the Aristotelian revival of the thirteenth century, however, these theories gradually assumed dominance but did not shed all their Hippocratic and Galenic elements.[31] By the later thirteenth century the simple incompatibilities between the two sets of theories had 'become blurred and complicated by a welter of distinctions and mutual accommodations'.[32] The result was to compromise Aristotle's notion of the impossibility of hermaphrodites and tacitly to admit their existence, but associating them henceforth with sexual duplicity or deviance,[33] a category less associated with anatomical difference and more closely attuned to sexual or gender deviance.

There was increased interest in hermaphroditism, to judge by the number of books published on the subject, from the early sixteenth century onwards. This was partly as a result of scientific developments in understanding nature and partly as a result of social changes altering perceptions of the 'proper' place of men and women.[34] Daston and Park argue that 'The range and intensity of this medical interest in the topic was distinctly new – in contrast to the relatively brief and general references in earlier treatises – as was the urgency of the moral and social concerns that they expressed.'[35] From the late sixteenth century, however, with a Hippocratic revival in full swing, medical texts began to associate hermaphroditism with sexually, morally and theologically charged issues of sodomy, transvestism and sexual transformation in a context where male power was being challenged.[36] The late sixteenth-century *On Monsters and Prodigies* by Ambroise Paré considered hermaphroditism alongside a number of 'monstrous' births and the book discussed several cases in which women changed into men. Daston and Park again: 'Rather than invoking Aristotelian considerations of excess maternal matter, only partially mastered by the paternal seed, [Paré] offers a frankly Hippocratic explanation', whereby the woman provided as much seed as the man with the result that two sexes may be found in the same body.[37]

Despite the title of Paré's work, a naturalist explanation of the genesis of the hermaphrodite was offered. Foucault and other historians have argued that from the late sixteenth and early seventeenth centuries the association between the hermaphrodite and the marvellous or monstrous began to decline as increasingly anatomical and biological accounts gained sway. Changes in the sex of individuals were deemed unlikely or impossible – any apparent change in sex was merely the coming to the fore of existing sexual characteristics. As the acceptance of the marvellous and monstrous declined so did the possibility of the real hermaphrodite.[38] Thus doctors who wrote about anatomy, such as the French anatomist Jean Riolan, began to argue in the early seventeenth century that the hermaphrodite did not – could not – really exist.[39] While this change in thought took some time and we should exercise caution in taking an isolated case or even a cluster of cases to presuppose the acceptance of a completely new paradigm, we can say that by the early nineteenth century old understandings of the hermaphrodite as a marvellous being or even as a natural possibility had well and truly faded.[40]

While this ongoing process eliminated the hermaphrodite as a category, remnants of this thought remained in scientists' interpretations of both the human and animal world throughout the period we study here. In nineteenth-century medical treatises much reference was made to Darwin's thought on the evolution of animals and plants with respect to the sexes and reproduction. Darwin was to write in his *Variation of Animals and Plants under Domestication* (1868), that 'in many, probably in all cases, the secondary characters of each sex lie dormant or latent in the opposite sex, ready to be evolved under peculiar circumstances'.[41] Such a theory of 'original bisexuality', whereby beings combined elements of the two sexes, one of which dominated the other before birth, would resurface, in, for example, Otto Weininger's *Sex and Character: An Investigation of Fundamental Principles* (1903). Weininger, like the Spanish scientist Gregorio Marañón later, held that the characteristics of the other sex never truly or completely disappear and that they '*persist without exception*'.[42]

While some, such as Paul Julius Möbius, understood sexually intermediate categories to be the result of degeneracy,[43] others understood 'bisexuality' to be natural. In Marañón's scheme, 'intersexuals' were those in whom the triumph of maleness or femaleness had not been sufficiently complete to entail proper 'sexual differentiation'.[44] Darwinian theories were also employed by psychiatrists such as the American James Kiernan who in 1884 understood homosexuality to be an atavism, a throw-back to lower evolutionary forms.[45] Physicians such as the Frenchman Julien Chevalier interpreted homosexuality as a faulty evolutionary development in 1893.[46] The British sexologist Henry Havelock Ellis cited a number of authorities such as Darwin, Haeckel and Kiernan to argue in his *Sexual Inversion* (1897) that homosexuality was a form of hermaphroditism, that is, a reversion to the primitive ancestral phase when bisexuality was the norm.[47] Eventually, however, the notion that hermaphroditism was the basis from which animal life sprang was slowly eliminated in the face of sexual dimorphism from the embryonic stage. Darwin's *Descent of Man* (1871) posited that while the progenitor of the vertebrate kingdom was androgynous, sex differences had come about in response to changing environmental conditions, which in turn impacted on evolution in tandem with natural selection in the form of sexual selection.[48]

Feminist epistemologies of the body: towards critical notions of gender, sex and the body

In 1903, Otto Weininger, the author of *Sex and Character*, asked himself and his readers the following question: 'where is sex situated and where is it not?'[49] His inquiry arose from his own interest in determining where masculinity and femininity 'express themselves', having made it clear that 'sexuality' (by which he understood maleness and femaleness) was not restricted merely to the reproductive organs and the 'sex glands' or gonads. Doctors who investigated cases of supposed hermaphroditism also asked themselves precisely where sex was located, in the body (and, if so, where exactly) or in the mind?

At least since Ann Oakley's *Sex, Gender and Society* (1972) and Gayle Rubin's analysis of what she termed the 'sex/gender' system (1975),[50] feminist thinkers have understood that the body is not separate from sex, that sex is not separate from gender and that these categories are products of historical human activity rather than 'real' in any transhistorical sense. While these accounts have often seen gender as a socially constructed set of signifiers and behaviours that become attached to the sexes, they have tended, at least until recently, to view the sex of the person as something intrinsic and biologically grounded. Gender, as a set of cultural practices, was understood to spring from the sex of the body within certain social circumstances.

Recent feminist authors, however, coincide in the need to revise some of the claims of 1970s feminism with respect to the body, sex and gender, to reassess the question of biology in action in society and the issue of how science and society construct nature and nurture. Nelly Oudshoorn discusses how during the second wave of feminism in the 1970s, male and female bodies were viewed from a perspective that rejected biological determinism (the position that posits essential differences between men and women in terms of bodies, hormones and psyches and believes that gendered social practices arise from an intrinsic sex-differentiated biological basis), focusing on the social as providing the constraints on (particularly) women's behaviour and abilities.[51] But feminism at this time did not enter into a critique of the notion of the 'natural body' or into an analysis of the power of the bio-medical sciences to proclaim truths about the body. Feminism, including the pioneering work of Ann Oakley mentioned above, focused instead on the social, following the argument held by Simone de Beauvoir that women are made and not born. This meant that many feminists accepted the distinction between innate biological sex differences and gender attributes acquired by socialization. Such a move effectively allowed feminism to regard the social as its point of research and debate, leaving the biological untouched as a category of reality outside of the social. As Oudshoorn states: '[T]he concept of sex maintained its status as an ahistorical attribute of the human body and the body remained excluded from feminist analysis.'[52]

There has, however, since the mid-1980s been a steady revision of the relationship between the social and biological sciences, of the concepts of nature/nurture and of the biological body.[53] Not least, this entailed a revision in feminist thought of the idea that sex and biology in themselves are fixed categories.[54] From a feminist perspective, such reassessments have resulted in considerable shifts in this area of thought. Schiebinger, for example, now notes that feminism has argued that gender differences are not fixed in the character of the species but 'arise from specific histories and from specific divisions of labour and power between the sexes'.[55] While feminists, Schiebinger continues, have opposed the argument from 'nature' with one from 'nurture' since the seventeenth century, recently two caveats have arisen: first, too strict a demarcation between nature and nurture can obscure how 'nurture' (culture) can form 'nature' (the body). Second, having accepted a strict division between nature and nurture feminists allowed a certain constructivism to prevail that tended to dissolve all body differences into political and cultural artefacts. Recent developments in medicine have shown how nature too needs to taken seriously with respect to women's health issues, for example.[56]

While the biology of the body, the constructions of nature and nurture and the materiality of the body have all recently come under scrutiny, some feminist biologists are curious that it is still mainly only the surface of the body that has received social analysis. Although the body's outside appearance can be adorned or physically altered to fit with changing cultural mores, for Birke 'the renewed focus seems always to end at the body's surface'.[57] Birke argues for a more profound social theory: 'While recent sociological and feminist theory has made enormously important claims about the processes of cultural inscription *on* the body, and about the cultural representation *of* the body, the body that appears in this new theory seems to be disembodied – or at the very least disembowelled. Theory it seems, is only skin deep.'[58] Birke proposes a look inside the body to see how assumptions about gender 'are read onto nature, including the insides of our bodies'.[59]

Judith Butler, for her part, critiques the feminist distinction between gender and sex as she understands this dualism as retaining a binarism between male and female, masculinity and femininity, reproducing dominant heterosexual social and sexual relations. Gender and sex, according to Butler, arise from performances which need to be reiterated continually in order to retain their unchallenged significance and hegemony in the social scene. Sex and gender are not seen, respectively, as categories that emerge from 'nature' and 'culture'. Rather, our understandings of both arise from our interpretation of nature; indeed, notions of gender, that is, the performance of gendered acts, Butler argues in her *Gender Trouble*, actually make up what sex we are.[60] As a result of the critique by some feminists and others that Butler's account viewed bodies and sexual identities as mere responses to discourse, without considering the social structures of society or materiality, Butler responded in her later *Bodies that Matter* that she understood that the materiality of the sexed body is itself socially constituted.[61] The sexed body would also form part of what Foucault termed a 'regulatory ideal' that produces the bodies it governs. Sex, for Butler, 'is a regulatory ideal whose materialization is compelled, and this materialization takes place (or fails to take place) through certain highly regulated practices. In other words, "sex" is an ideal construct which is forcibly materialized through time.'[62] As such, 'sex' is materialized through the reiteration of the norms that govern it.[63]

While Butler theorizes 'performativity' as the means by which sexed bodies, gender and sexuality are constituted socially and materially, some feminist authors remain convinced that gender is still useful as a category of analysis. While 'gender must be undone',[64] deconstructed and exposed as a historical and social construction, Iris Young, for instance, argues that Judith Butler's and Toril Moi's recent accounts, despite their differences, focus more on the constitution of subjectivity and identity and less on social structures and processes of discrimination and oppression.[65] What Young proposes is the use of gender as a category of analysis which is socially and materially grounded: 'Gender ... is best understood as a particular form of the social positioning of lived bodies in relation to one

Introduction 13

another within historically and socially specific institutions and processes that have material effects on the environment in which people act and reproduce relations of power and privilege among them.'[66]

For the purposes of our account of the science of hermaphroditism here, we attempt to do justice to an analysis which tries to proceed beyond the binarisms sex/gender, nature/culture, male/female, not as an ungrounded utopian move, but as one that shows how sexed identities, bodies and sexualities were constructed and disciplined in a matrix of unequal power relations at a particular historical time. Hermaphroditism, we will argue, can be analysed historically by examining how sex, gender and the body have been both discursively and materially constructed – how they have been evoked and brought into being, so to speak – at different times over the last centuries. We will see, by means of a number of closely read examples from Spain during the years 1850–1960, how sex, gender and the body were minutely policed categories, brought into existence and sustained by systematic reiteration of their qualities. We will examine how these categories actually produced each other in a complex process of 'sex determination' by specialists. Sex, gender, the body and sexual orientation 'are expected to be concordant in each individual' and once an observer has attributed 'male or female sex to a particular individual [he or she] assumes the values of other unobserved characteristics'.[67] On doing so, we will see how useful these categories are for successful historical analysis. Again, we do not wish to posit the firm existence of these categories and then read them back onto historical scenarios. For example, we will argue that 'gender' is a useful category of analysis from the mid nineteenth century onwards,[68] as the sexes became more dichotomous, but this is not necessarily so for the sixteenth century (discussed in Chapter 2) when sex differences were marked less by supposed intrinsic differences between bodies and more so by social and political differences such as rank, economic activity and dress.[69]

Chronologies of the science of hermaphroditism: from physical touch to psychological gender tests, 1850–1960

We have noted that the notion of the hermaphrodite as a marvellous entity continued in the western imagination amongst natural philosophers and literary sources until the early seventeenth century despite certain changes in meaning. During the first decades of the 1600s a number of texts would significantly revise perceptions of the hermaphrodite. The 'real' human hermaphrodite, uniting more than one sex in one body, slowly lost currency and by the end of the nineteenth century was deemed virtually impossible. If texts from the 1600s declared the hermaphrodite a questionable entity, it was during the early 1800s that doctors presented with cases of ambiguous sex began to draw up taxonomies for the identification of a person's real biological sex, often for legal purposes or for the granting or confirmation of permission to marry. Instead of classifying most of these persons as real hermaphrodites, doctors detailed and graded the differences between the sexes. Certain differences between the sexes could be combined in the same body, but doctors would determine precisely which characteristics predominated in order to permit a definitive classification of the person according to maleness or femaleness, apparently in accordance with the two-sex model. Those individuals that displayed a mixture of characteristics would not be labelled real hermaphrodites; classificatory systems allowed for a new category: the *apparent* hermaphrodite.

One of the first classificatory systems to detail these differences was that of Charles Chrétien Henri Marc (1771–1840). In 1817 Marc emphasized what he called 'vices of conformation' in human beings, which made identification of sex a somewhat problematic endeavour.[70] While Marc rejected the existence of 'real' hermaphroditism in 'perfect animals', a category that included human beings, he did allow for cases of apparent hermaphroditism in both the female and male sex. He envisaged, however, a further category: 'des individus sans sexe bien déterminé et que l'on ne peut, par conséquent, considérer comme mâles ni comme femelles'.[71] These would be 'neutral' hermaphrodites or, after Quintilian, *genus epicoenum*, consisting of very few cases. We are clearly at an intermediary stage between the full elimination of the category of the hermaphrodite and the acceptance that a small number of individuals could fall into this 'neutral' position. Steadily, even this third category would be eliminated as the nineteenth century wore on.

As part of his classificatory system Marc laid down a series of stages in the examination of an individual of indeterminate sex. These steps included the physical countenance of the individual, an account of his/her likes and dislikes and of any menstrual activity, and the consideration of any declarations made by the 'hermaphrodite' him- or herself. The relative importance of these constituent elements would evolve over time. From the early nineteenth to the mid-twentieth century the anatomy of the person tended to be regarded as the principal means of sex identification, even though precisely which bodily parts were deemed to be important altered over time. By the 1950s, however, so-called secondary and tertiary factors and the perception of the 'patient' him- or herself were major, if not deciding factors as 'transsexualism' became a gendered and sexual category.

Marc's early nineteenth-century system was elaborated upon by other anatomists such as Isidore Geoffroy Saint-Hilaire (1805–61) who, writing in 1833, classified hermaphrodites quantitatively, where excess in either female or male parts determined the type of hermaphrodite. James Young Simpson (1811–70) in the late 1830s suggested that *apparent* sexual traits allowed for a division between 'true' and 'spurious' hermaphrodites. It was only in 1876 that Theodore Klebs divided hermaphrodites into 'true' and 'false' or pseudo-hermaphrodites on the basis of an analysis of the sex glands or gonads of the individual.[72] A further classificatory system was drawn up by Jean Samuel Pozzi who wrote of 'androgynoid' hermaphrodites, women who looked like men, and 'gynandroids', that is, men who looked like women. They were not real hermaphrodites but men or women who looked like the other sex.

What methods of diagnosis allowed doctors to identify the real sex of the body? If previous methods had relied primarily upon sight and touch as the principal tools for sex identification by examining the external genitalia, from the early to mid-nineteenth century the criteria aiding identification changed to incorporate a broader variety of characteristics including comportment, hirsuteness, timbre of voice and existence of breasts and gestures, as we have seen with Marc's descriptions. From the late nineteenth century, new diagnostic techniques allowed for the extirpation and analysis of the ovaries and testes under the microscope.[73] For some time doctors had entertained doubts about *ante-mortem* diagnosis as on several occasions the *post-mortem* analysis gave different results as regards the sex of the individual.[74] In this way, the ovaries and the testes, or the gonads, became, according to Dreger, the overriding factor in the determination of sex as part of a process of medicalization of sexuality unprecedented in Europe.[75] Further, once samples of the ovaries or testes could be extracted from living bodies with techniques such as exploratory surgery or laparotomy, doctors believed they could pronounce on the sex of the person with more certainty by means of histological techniques.[76]

The science of hermaphroditism, however, did not rely on sex identification from the gonads alone for long. From the 1920s, with the discovery of 'male' and 'female' hormones, endocrinological accounts of real sex took centre stage. The 'truth' of the body was now found to lie in the hormones secreted by the 'sex glands' and not necessarily in the tissue of the gonads. However, in the late 1920s it was confirmed that some hormones previously known as 'female' were found in non-hermaphroditic males and vice versa. The idea that the two sexes were radically different and incompatible led, with the aid of new psychological theories of gender development, to the notion that the sexes were graded rather than rigidly differentiated and that a variety of factors – genital, gonadal, hormonal and psychological – conferred the real sex of the person. In a sense, modern medical science, as a confluence of anatomical, endocrinological and psychological techniques, had returned to its Hippocratic roots, whereby sex was seen as a continuity, a number of gradations or intermediate positions between female and male. In the final section of this introductory chapter we turn to the current state of research on the hermaphrodite in Spain and comment on the validity of the above schema.

Spain: current state of research

The European legacy of thought on the generation of human beings and hermaphrodites was, as we have seen above, a complex one. So much is true with respect to medieval notions of the sexes and the development of a 'science of hermaphroditism' in the eighteenth and nineteenth centuries. While we cannot always claim that Spain was exceptional in terms of the history of ideas in general and the history of sexuality in particular, it is necessary to exercise a certain degree of caution when accepting accounts elaborated on the basis of other countries' data. Elsewhere, we have warned against the transposition of a 'vulgar' understanding of Foucault's work on the history of homosexuality, which would seek to plot the French author's insights on the subject, largely elaborated from French and German models, on to other countries.[77]

Despite these initial caveats, it is possible to argue that in Spain, as in other European countries, there was a combination of Aristotelian and Hippocratic understandings of the generation of humans in general and of the hermaphrodite in particular. Medical authors such as Fragoso (1570) and Alfonso Carranza (1630), authors of books on marvels including Antonio de Torquemada (1570) and Antonio de Fuentelapeña (1676) as well as other authors (for example, Huarte de San Juan, 1575; Covarrubias, 1611) all coincide on the existence of hermaphrodites as real beings and accept that sex changes take place as natural events, responding to what would appear to be a dominant Hippocratic-Galenic interpretation.[78]

There are, however, as in other European countries at the time, some exceptions to this Hippocratic dominance or idiosyncratic combinations of Hippocratic and Aristotelian positions. The above-mentioned Antonio de Torquemada's *Jardín de flores curiosas* (1570) cites Aristotle's *De Generatione Animalium* and interprets the latter as providing evidence for sexually intermediate figures. Pedro García Carrero (*Disputationes Medicae Super Libros Galeni*, 1605) places hermaphrodites in the category of monsters, apparently in an Aristotelian framework.[79] Gaspar Bravo de Sobremonte (*Operus Medicinalium Tomus Quartus: Tres Disputationes Complectens*, 1679) argues that males and females are formed differently and that no sex changes can take place.[80] This author does not, however, argue that two distinct sexes, male and female, actually exist; we are not yet in the 'two-sex' model period. Instead, he argues that any supposed changes in sex from female to male that do take place are in reality women with overlarge members or hermaphrodites that possess both sexes. This eclectic legacy, with apparently no one dominant understanding on the generation of human beings or the establishment of the sexes, gives rise in Spain to a similar situation to that of other European countries, if the analysis by Daston and Park is correct.[81] Such a state of affairs continued to inform Spanish science well into the twentieth century.

Existing research on hermaphroditism in Spain has been largely confined to the sixteenth and eighteenth century with only one study on the contemporary period.[82] This study seeks to analyse how discourse on hermaphroditism may be recast by means of a strong comparative approach. Spain, despite following broad European trends in sexual science, is posited not as a country that merely reproduced others' ideas about hermaphroditism but, from a dynamic and comparative methodological perspective, one in which scientists actively engaged with imported theories, elaborating local understandings of their own making. In the light of this guiding framework and the paucity of studies on hermaphroditism in Spain to date, this study sets out a number of objectives that are translated into chapter divisions. In Chapter 2 we aim to trace the process by which the hermaphrodite lost its marvellous or monstrous status. We emphasize the uncomfortable juxtaposition or fusion between principally Hippocratic understandings of the generation of the sexes and those more indebted to an Aristotelian framework. This allows us to chart through a close discussion of texts from the sixteenth to the eighteenth century how the possibility of hermaphroditism was gradually eliminated in favour of the acceptance of pseudo- or apparent hermaphroditism within a changing and increasingly technological medicalized environment and within a dominant acceptance of the 'two-sex' model.

If the individual in the nineteenth century became, according to Foucault, an effect of power in the context of the construction of the sexual sciences and the medico-legal profession's interest in the identity of the individual, he or she was also brought into the limelight by a cluster of social concerns.[83] Anxieties over the lack of virility which had allowed Spain's last colonies to disappear, as the country lunged into a deep crisis around and after 1898, the growing contestation of women's movements, an alleged crisis in the birth rate, the acknowledgement of the 'social question' and powerful, destabilizing working-class movements, all placed emphasis on the need to seek out pathological and dissident strains in the national body.[84] One effect of the category of individuality that liberal philosophy had bestowed on its subjects was the juxtaposed demand for greater rights and the dangers of contamination by individuals and social groups previously unknown or rarely acknowledged.[85] The hermaphrodite, like the alcoholic, the homosexual and the criminal, posed a threat to this emerging liberal order, a threat that had to be contained and managed.[86] There were two main ways in which those of doubtful sex inhered in this *mélange* of concerns: marriage legislation and recruitment for the army. In the first instance, there was no legislation specifically in Spain on the subject of hermaphroditism but medical doctors could pronounce a marriage null and void if both partners were found to be of the same sex or if supposed hermaphroditism gave rise to incurable infertility. One consequence of the ongoing colonial wars of the nineteenth century was regular call-ups to the army. The legal doctor's duty was to certify the maleness of the conscript.[87] Chapter 3 begins with these concerns in the context of the arrival of the age of the gonads, as Dreger has termed it, from the 1870s.[88] From this time, according to the same author, hermaphrodites were understood to have been assigned the 'wrong' gonads in relation to their sex and it was the duty of medicine to correct this anomaly.[89] Dreger and Oudshoorn argue that this philosophy continued to hold sway up to the mid-1920s when it fell into crisis. This chapter, while fundamentally agreeing that the importance of the gonads in sex determination was dominant during this period, also argues that it was not necessarily as hegemonic as these authors suppose, at least in Spain. Instead of a fundamental and prevailing acceptance of the gonadal model, in fact what we see is vacillation and eclecticism on the part of doctors. Nevertheless, Chapter 3 sets the scene for the emergence of other frameworks, those of endocrinology first and psychological gender science second, which are both discussed in Chapter 4.

Chapter 4, covering the period 1920–39, discusses what can probably be held as the most productive period in the science of hermaphroditism given the number of explicit cases analysed in the annals of legal medicine. Again, through a close analysis of these cases, a picture is drawn up whereby the relative importance of anatomical, gonadal and, later, endocrinological factors in the genesis of hermaphroditism is discussed. Through an analysis of these texts, once again, we emphasize the impression not of a science that is sure of itself but one that is very much in the making. The deliberations and doubts of medical figures are often recorded in the pages of the prestigious medical reviews which record these cases.

Internationally, and particularly in the United States, the constant innovations, renovations and crises of hermaphrodite science would eventually allow for the emergence of a cluster of psychological theories on 'gender identity', that is, one's sense of oneself as male or female, distinct from 'gender role', the cultural expectations according to maleness or femaleness.[90] 'Gender', as distinct from sex, had been theorized first in the States by doctors such as John Money, Joan Hampson and others, and in the 1950s it became a key concept when analysing cases of hermaphroditism.[91] While Spanish doctors did not yet use the term 'género', we note a marked concession being made to the patient's perceptions of him- or herself and their 'sexual orientation', another concept not yet coined. The growth and influence of these theories, now under the mantle of the Franco dictatorship, are assessed in Chapter 5.

This major shift began to theorize more closely what doctors understood to be the misalignment of gender and the body, paving the way towards the transsexual, a person who would *become* the other sex by means of psychological and medical intervention. It is here that we leave the trail of the 'hermaphrodite' in Spain, making some final remarks on our study in the Conclusion and signalling future venues for research for the post-1960s period.

Throughout this book we acknowledge that science and medicine are not conducted in isolation from the society in which they are performed. Scientists, as the history and sociology of science have clearly shown, imbue their research programmes and their findings with cultural notions that in turn structure scientific thought.[92] We are keen to examine the extent to which an apparently growing interest in hermaphroditism, in tandem with changing medical discourse on hermaphroditism, reflected social anxieties about shifting gender roles, the evolving discourse on sexuality and, in particular, the increased visibility of the 'sexual deviancies' such as homosexuality and changing legislation on marriage and divorce.

One of the major concerns of doctors dealing with cases of sexual ambiguity in the late nineteenth century and up to at least the mid-twentieth was the elimination of the discursive and actual possibility of 'homosexual marriage' through wrongly assigned sex or the approval of same-sex activity as a result of any investigation into a person's 'real' sex. On this point, Suzanne Kessler notes 'As recently as 1955, there was some concern that if people with the same chromosomes or gonads paired off, even if they had different genitals, that might bring the physician in conflict with the law for abetting the pursuit of (technically) illegal sex practices.'[93] In the light of this appreciation, and indeed in general terms, we wish to assess whether the ideology of the Franco dictatorship from 1939 onwards had any effect on the articulation or the constraints of the science of hermaphroditism.

Introduction 21

We have also tried to engage with insights on sex differences and the hermaphrodite from the broader sexological and medical press and from the literary world. We take it as axiomatic that popular perceptions of sexual issues would not necessarily follow scientific ones in chronological terms or with respect to actual perceptions. Literary representations coexist with scientific ideas in an often highly productive equation and we attempt here to evince chronological and conceptual disparities of hermaphrodites between medicine and more popular accounts.

We also note that, despite sixteenth-century and early nineteenth-century conceptualizations that steadily eliminated the hermaphrodite from the range of human possibilities, doctors in their accounts of intersexual figures often slipped between 'pseudo-hermaphroditism' and 'hermaphroditism'. For example, in *España Médica* in 1860 a case of 'hermaphroditism' was discussed;[94] likewise in *La Medicina Ibera* in 1920;[95] in 1924 a case of 'pseudo-hermaphroditism' was discussed in the same journal;[96] two case of 'real hermaphroditism' were analysed in 1956.[97] Several cases of 'pseudo-hermaphroditism' were discussed in the 1960s and others denoted 'hermaphroditism'.[98] While there are differences between these cases, as we shall see in Chapter 5, we believe this rather inconsistent terminology responds to an unwillingness to jettison the age-old term 'hermaphroditism' even though medical diagnosis generally recognized that such a condition was extremely rare or nigh impossible.

Finally, it is our hope that the voice of those whom medicine deemed hermaphrodites can to some degree be heard in this study in order to examine their own self-concepts and their internalization of or resistance to medical knowledge about their 'condition' as part of an exercise in medical history 'from below'.[99] There is a historiographical question mark over the extent to which doctors designated the sex of their patients (and perhaps operated on them) without their consent and the extent to which medical doctors accepted that the person in question should live their life as the sex they felt most comfortable with. This question provoked heated discussion amongst doctors themselves and sometimes international controversies.[100] Often, however, doctors dispensed with the patient's voice in their medical reports. We hope, in contrast, that our account will make a contribution to making the voice of the patient heard as part of a process of empowerment in the face of the challenges provided by the early twenty-first century.

Notes

1. Ovid (2004: 144–50).
2. Ibid., p. 150.
3. Ibid. See also Dreger (1998: 31–2); Brisson (2002).
4. On the androgyne as a feminist ideal, a way of developing love relations between equals, as a model of beauty, often with a certain homoerotic element, and as myth see, respectively, Ferguson (1991: 189–216); Weil (1992); Mosse (1994: 259–60). For the Spanish context see Frattale (1989: 9–42).
5. See Daston (1991).
6. See Park and Nye (1991: 53); Paré (1982: 31–3).
7. Vázquez García and Moreno Mengíbar (1997a: 191–6).
8. Foucault (1980).
9. See Fausto-Sterling (1997: 219–25).
10. Labanyi (2000); Ortiz (1993).
11. See Epstein (1990: 101–4).
12. We view the process of the medicalization of hermaphroditism as analogous in many ways to that of homosexuality in the late nineteenth and early twentieth centuries. See Cleminson and Vázquez García (2007).
13. As argued in Foucault (1990a: 43) where he notes 'Homosexuality appeared as one of the forms of sexuality when it was transposed from the practice of sodomy onto a kind of interior androgyny, a hermaphroditism of the soul.'
14. On occasion the categories of homosexuality and hermaphroditism came together although, as we have said, the tendency was for the elimination of same-sex desire. For 'homosexual hermaphrodites' see Dreger (1998: 126–30). For a classic text in this sense see Magnan and Pozzi (1911).
15. Dreger (1998: 31) notes that it was Richard Goldschmidt (1917) who apparently first used the term 'intersexuality' to denote a wide range of sexual ambiguities including what had previously been known as hermaphroditism. For the invention of the term 'psychopathia transexualis' by Dr David O. Cauldwell, see Meyerowitz (2002: 42).
16. While a similar critique has often been applied to histories of male and female homosexuality, and has been represented in terms of a struggle between 'essentialism' and 'social constructionism' since the late 1970s, it is only more recently that such debates have inhered in the history of hermaphroditism, transsexualism and transgenderism. See the questioning of the use of the term 'transsexuals' to describe the cross-dressing Omani *xanith* by Garber (1991: 239–45).
17. See Dreger (1998); Oudshoorn (1994).
18. There are other influences or channels of thought on the subject that contributed to the consideration of hermaphrodites in the West, such as Pliny's account of them, following Aristotle's discussion, in Pliny (1942: 517 and 529) where he notes that hermaphrodites were now considered to be entertainments (529).

19 See, for example, Boylan (1984); Cadden (1993); Park (1985).
20 Plato wrote in his *Symposium* that human beings were originally divided into three sexes, male, female and androgynous, until all types were split into two in order to punish them for their refusal to pay homage to the gods. In Plato's account, human beings would forever wander the earth in an attempt to find and reunite themselves with their other half; of the various recombinations 'heterosexuality' and 'homosexuality' would result. See Aristophanes' exposition of love along these lines in Bloom and Benardete (2001: 18–22).
21 See Dean-Jones (1994: 166–70). Dean-Jones notes that the origin of the male seed was important. If it came from the left testicle it was more likely that a female would result; if it came from the right testicle a male was more likely (167).
22 Laqueur (1990); Moscucci (1991: 175).
23 Moscucci (1991: 177).
24 Cadden (1993: 3).
25 Daston and Park (1996: 118). Other examples of a critique of Laqueur's premises of Hippocratic dominance include Park and Nye (1991) and Stolberg (2003) where the dates around which the 'one-sex' model gave way to the 'two-sex' model are contested. See also the responses to Stolberg by Laqueur (2003) and Schiebinger (2003).
26 The complex legacy of Aristotle's thought and the differences between the Aristotle of *De Generatione Animalium* and his later works, which generated confusion in the ensuing centuries, for example in the work of the physician Galen (131–201 CE), are examined in Boylan (1984). Aristotle, in contrast to what Pliny reports (Pliny, 1942: 517), did not appear to countenance human hermaphrodites, at least in *De Generatione Animalium*, but did write of animals having too few or too many parts; in the latter category *tragainai*, hermaphrodite goats, would fit. See Aristotle (1943: 427).
27 Dreger (1998: 32).
28 Ibid.
29 This paragraph follows Cadden (1993: 33–4).
30 Daston and Park (1996: 120), citing Cadden (1993: 39–110). The influence of Arabic thought on the question of hermaphroditism, particularly with respect to Spain, is in need of further research. Greek philosophical thought was 'rediscovered' via the Islamic presence in Italy and Spain in the early new millennium, translated from Arabic into Latin and eventually vernacular languages. See Jacquart (1997). The Iberian wave of this thought via Arabic sources came in the twelfth and thirteenth centuries when there was a popularization of Aristotle's thought with Abd al Malik Ibn Zuhr's medical treatise and the *Generalities* (1194) of Ibn Rushd (Averroes), who discusses Aristotle and Galen. See Fletcher (1998: 8–9; 133). Spink and Lewis (1973: p. ix), in the 'Introduction' to Albucasis' work *On Surgery and Instruments*, note that the fame of the *Surgery* spread in the Islamic world and in the second half of the twelfth century it was translated into Latin at Toledo by Gerard of Cremona as *Liber Alsaharavi de*

Cirugia. In Chapter 70, 'On the treatment of the hermaphrodite', three kinds of hermaphrodite are discussed: two male and one female. In these, 'the superfluous growths must be cut away so that every trace is destroyed; then the usual treatment for wounds should be applied' (454). For the kind of male hermaphrodite who has an opening in the scrotum from which the urine runs, there is no cure. On the one hand, Albucasis appears to follow Aristotle's notion of hermaphroditism as excess flesh but on the other admits that hermaphroditism does exist.

31 Daston and Park (1996: 120) following Cadden (1993: 53–104) and Baldwin (1994: 94–6).
32 Daston and Park (1996: 121).
33 Ibid. The authors point out that this idea was not necessarily carried forward into medical texts, which avoided moral judgements or left them to theology (as in the case of sodomy).
34 Oakley (1972: 9) argues that differences between the sexes become most hotly debated in periods when 'the existing roles and status of male and female are changing'. One of those periods, at least in England, would be the century from 1540 to 1640. A degree of equality had been achieved by the mid-sixteenth century but for some this equality was seen as a threat to the prescribed social role of the sexes.
35 Daston and Park (1996: 118).
36 As argued in Daston and Park (1996: 122) and Oakley (1972: 9). We should not underestimate, however, the changes that took place on the question of homosexuality during previous periods. According to Boswell (1980: 269–332), it was during the thirteenth and fourteenth centuries that social and theological condemnation of homosexuality gathered strength. Medical and legal texts, however, may have delayed their censure. On this point, see Foucault (1994: 293–4).
37 Daston and Park (1996: 121). This paragraph follows Daston and Park's analysis.
38 See Foucault (2001: *passim*) and Vázquez García (1999).
39 Jean Riolan was one of many seventeenth-century anatomists who explored sex differences. Riolan's *Discours sur les hermaphrodits, où il est démontré, contre l'opinion commune, qu'il n'y a pas de vrais hermaphrodits* (Paris, 1610) is discussed by Foucault in *Los anormales* (2001: 71–4), and his *Anatome* (Paris, 1610) is mentioned in Stolberg (2003: 281 and 284), as evidence of a growing referral to a two-sex model.
40 On the danger of judging 'foundational shifts in scientific culture merely by firsts' see Schiebinger (2003: 307).
41 Cited in Meyerowitz (2002: 22–3). Moscucci (1991: 178) has argued that 'The nineteenth-century fascination, one might even say obsession, with the latent hermaphroditism of humankind can be seen as an attempt to reconcile the concept of sexual difference with the idea of a human nature common to both sexes.'
42 Weininger (2005: 12). Emphasis in original.
43 Steuer (2005: p. xx).

44 See Marañón (1972).
45 See Herrn (1995: 37).
46 Ibid. On Chevalier's evolutionary ideas with respect to human intellect and sexuality see Rosario (1997: 87–8).
47 As discussed in Carter (1997: 161 and 164). Geddes and Thomson, the authors of *The Evolution of Sex* (rev. edn 1901), viewed hermaphroditism as a primitive mechanism of reproduction confined to the 'lower' species while in the higher groups sexual dimorphism was a marker of advanced development. See Russett (1989: 135–6). Nye (1989: 41, n. 35) notes that late nineteenth-century French doctors would use the terms pederasty, inversion and psychic hermaphroditism as overlapping phenomena, whereby the last two would be typical qualities of the pederast. He cites Alexandre Lacassagne, 'Pédérastie', *Dictionnaire encyclopédique des sciences médicales* (Paris, 1886) to this effect. Psychic hermaphroditism would be, once again, not 'real' anatomical hermaphroditism but the belief that one was in reality the other sex or one was acting out the other sex's behaviour.
48 Moscucci (1991: 182–3).
49 Weininger (2005: 16).
50 Oakley (1972); Rubin (1975: 157–210).
51 Oudshoorn (1994: 1–2).
52 Ibid., p. 2.
53 The literature on the body is now extensive. See Connell (2001) for an overview of recent developments in the sociology of the body.
54 Hausman (1995: 8–9).
55 Schiebinger (2000: 1).
56 Ibid., pp. 2–3. Earlier accounts include Mauss (1992) where he notes the different types of bodily activity and the influence of what he terms the 'habitus', as collective habits and customs, to describe body practices.
57 Birke (1999: 2).
58 Ibid.
59 Ibid., p. 7.
60 Butler's position is not dissimilar to that sustained by some Spanish feminist analysis. The concept of 'discurso genérico o sexuado' (gendered or sexualized discourse) posits an interdependent process of construction of 'gender' and 'sex' as proposed in Valcárcel (1991).
61 We follow the account given by Young (2005: 14–15).
62 Butler (1993: 1).
63 Ibid., p. 2.
64 Butler (2004).
65 See Young (2005: 12–26).
66 Ibid., p. 22.
67 Chase (1998: 190).
68 See Scott (1999: 28–50).
69 In this sense, we have some difficulties in completely endorsing Laqueur's perspectives on sex and gender in his *Making Sex* (1990). On the one hand, he states that sex, or the body, must be understood

as the epiphenomenon, while gender would be the primary or 'real' category to distinguish the sexes (p. 8). Two genders, in this reading, would correspond to one sex (p. 25). But Laqueur seems to underestimate the materiality of the body and how differences such as activity/passivity, warmth/coolness and dryness and moistness are imprinted on the body with social effects. He also states that 'So-called biological sex does not provide a solid foundation for the cultural category of gender, but constantly threatens to subvert it' (p. 124) but also argues that the 'modern question, about the "real" sex of a person, made no sense in this period, not because two sexes were mixed but because there was only one to pick from' (p. 124). But we would also argue that, for the period referred to here by Laqueur, it makes no sense either to talk about 'gender', a category that only emerges in contradistinction to 'sex' from the eighteenth or nineteenth century onwards. Some suggested routes out of this conundrum are offered in Chapter 2.

70 Marc (1817). The notion of vices of conformation was taken up notably by the French forensic doctor Ambroise Tardieu in his discussion of hermaphroditism and homosexuality in the 1860s. On the impact of Tardieu and his thought generally see Rosario (1997: 72–7). On Tardieu in Spain see Huertas García-Alejo (1990: 91–4).
71 Marc (1817: 89).
72 For more details of these systems of classification see Dreger (1998: 140–7).
73 Dreger (1997: 46–66) puts some of the late nineteenth-century interest in hermaphroditism down to developments in anatomico-pathological research. On these developments, see Ackerknecht (1967), Foucault (1989: 124–48), and, Laqueur (1990: 70–96).
74 See Dreger (1998: 86).
75 For some views on the medicalization of society, see Foucault (1989). See also the analysis by Rose (1994).
76 Dreger (1998: 93 and 146–50).
77 Cleminson and Vázquez García (2007). For an essay that urges a close examination of national sexological traditions and a questioning of Foucault's discontinuous historical analysis, see Nye (1991). A further national difference with respect to hermaphroditism has been discussed by Dreger in that French doctors were keener than British ones to inquire into their patients' sexual penchants and activities, perhaps in order to unmask any possibility of practising homosexuality. See Dreger (1998: 50).
78 See Fragoso (1570); Carranza (1630); De Torquemada (1943); De Fuentelapeña (1978); Huarte de San Juan (1977); Covarrubias (1979: 531).
79 García Carrero (1605).
80 Bravo de Sobremonte (1679).
81 Daston and Park (1996). These authors also point to the idea (131, n. 13) that often the strongest Hippocratic position was taken by

more marginal members of the medical profession who used it to contest traditional medical authority, for example, Paré and Duval. They state that probably for this reason, a Hippocratic position is more common in vernacular writing than in Latin in the late sixteenth century. While the language relation outlined here shows a correlation with Spain, that is, that Hippocratic understandings were expressed in the vernacular, it cannot be said necessarily that this was articulated by the more marginal in the medical profession.

82 See Vázquez García and Moreno Mengíbar (1995b, 1997b); Vázquez García (1995, 1999); De la Pascua Sánchez (2004) where the case of the nun Fernanda's transformation into Fernando in 1792 is discussed; Cleminson and Medina Doménech (2004). A recent much more medical account is Salamanca Ballesteros (2007).
83 Foucault (1980).
84 See Cleminson and Vázquez García (2007: 175–215).
85 Vázquez García and Moreno Mengíbar (1997a: 247–9).
86 In France, vigilance in places of disciplinary control was also deemed important not just for the presence of inverts but also hermaphrodites. The 'wolf in sheep's clothing' scenario allowing for dangerous *de facto* heterosexual relations or unwitting homosexual relations in such places is discussed in Dreger (1997: 50–2).
87 Vázquez García and Moreno Mengíbar (1997a: 204 and 211).
88 Dreger (1998: 139–166).
89 Dreger (1997: 53–4) notes the concern amongst French physicians to unmask cases of same-sex marriage and to determine examples of 'erreur de sexe' in a much more explicit way than, for example, British doctors. In general, in France, such errors were announced to the patient; in Britain it was kept more confidential, and the patient may have been operated upon being ignorant of her 'true' sex.
90 Kessler notes that historically psychoanalysis tended to blur the distinction between role and identity by equating a person's acceptance of his/her genitals with gender role and ignoring gender identity. Freudian theory posited that if one had a penis and accepted its reality, masculine gender role behaviour would follow. See Kessler (1998: 14, n. 9). Freud, however, refused to accept any direct link between bodily 'feminization' of the male hermaphrodite and the homosexual's sexual preference, separating out the 'mental construction' of the homosexual from his or her physical form, as discussed by Gilman (1994: 335–6).
91 Money (1955); Hampson (1955); Money et al. (1955), where it is argued that gender identity is changeable until approximately 18 months of age.
92 See, for example, Latour and Woolgar (1986); Barnes et al. (1996).
93 Kessler (1998: 27).
94 Alba y López (1860). See Vázquez García and Moreno Mengíbar (1997a: 211).
95 Anon. (1920a).
96 Cardenal. (1924).

97 Botella Lluisá (1956).
98 See, for example, Fontes Gil et al. (1960); Cañadell and Planas Guasch (1961).
99 An excellent example of this approach with respect to homosexuality is Oosterhuis (2000). Karkazis (2008) has integrated the lived experience of intersex people into her more present-day study. See also Roy Porter (1985) on taking account of the patient's view in the construction of medical discourse and practice.
100 An exploration of the self-image of the hermaphrodite is found in the recent article by Mak (2005). Interestingly, Mak discusses cases where the doctors take into account the desire of the patient to remain one sex or the other at the beginning of the twentieth century as a minority or new phenomenon at that time. However, a similar case was discussed in Spain in the nineteenth century with respect to whether the operation performed by the surgeon Dr Ulibarri was one of 'complacencia' with the patient's wishes or one of 'necesidad' (Alba y López, 1861).

Chapter 2
From Sex as Social Status to Biological Sex

Introduction

Since Freud we have known that it is the subconscious that makes history. For this reason, just as the psychoanalyst must delve into personality conflicts, the historian is obliged to sift through the tacit truths that make up a culture in order to locate the 'lapsus' that functions as a revealing symptom.[1] This is what happens when the question of personal identity is examined and, more specifically, when sexual identity is examined in societies of high modernity. The distinction between men and women, understood as a given in the daily life of the *ancien régime*, only became problematic when cracks and tensions emerged around it. During this period, some individuals were pronounced hermaphrodites, there were cases of changing sex or people adopted the appearance of the other sex and these are documented extensively in medical, juridical, theological and literary sources of the sixteenth and seventeenth centuries. These examples reveal clearly how society at that time dealt with the question of sexual identity. In Spain, despite the fact that research has not been as extensive as for some north European countries, it is no longer possible to say that 'sexually intermediate' individuals have been without their historian.[2]

Any examination of the practices and representations which occurred around sexually ambiguous persons in modern Spain must necessarily draw on the work of Michel Foucault and Thomas Laqueur. Laqueur, especially, has marked out the terrain of historiographical analysis on this front in the last fifteen years.[3] Both authors have shown not only that sexual dimorphism (the identification of women and men as possessing distinct and incommensurable biological differences) is a historical construction of relative recentness. At the same time, they have separated gender characteristics (masculine/feminine), the basis of androcentric domination, from the biological dualism of the sexes (male/female). Gender divisions do not necessarily depend intellectually on sex differentiation. The idea that the distinction between men and women is based on biological differences between the sexes would be a cultural understanding consolidated by medical knowledge in the eighteenth century, thus breaking with the predominance of the hierarchical Hippocratic model of 'one sex' that was up to then hegemonic. This cultural understanding was also brought about, as Laqueur has argued, by the the new liberal political order or by the emergence of modern biopolitics and the nineteenth-century desire to determine the 'true sex' in cases of doubt (Foucault).

This is not the place to recount the details of this polemic, which, especially in the case of the discussion of Laqueur's work, is far from exhausted.[4] In this chapter we wish to signal the understandings afforded by this debate, as well as the obstacles that have emerged around it, in order to discuss the question of sexual identity in the modern age in Spain.

It is undeniable that the work of both Laqueur and Foucault on the sexed body has supposed a form of analysis that has left behind the naturalist focus offered by the biological sciences. Like any other object of sociological analysis, the sexed body must be understood in the context of a strict historical analysis.[5] The task for the social scientist and historian is to reconstruct, through comparison, the series of historical configurations, structures and differences inhabited by the subject. In other words, any analysis from a social science perspective should always locate the body in precise spatial and chronological frameworks.

Too often in the history of the biological sciences historical sources are interrogated as if they existed atemporally, to be understood with reference to existing legal frameworks and scientific terminology.[6] Such a methodology eliminates precisely what is of interest to the historian, who comprehends the body within a specific contextual and cultural context in which the body makes and is made up by predominating practices in any given society.

Foucault's and Laqueur's work on the history of sexual dimorphism has been fundamental for the understanding of the body as the subject of sociological analysis. The work of the sociologist does not rely on the discussion of empirical propositions, that is, falsifiable affirmations such as an examination of the 'errors' made by Renaissance anatomists on questions of 'menstruating men' or sudden changes of sex. Historians and social scientists deal with performative questions, not those of 'truthfulness' or 'falsehood'; they are interested in the 'grammatical propositions' of Wittgenstein and the 'énoncés' of Foucault.[7] This tactic does not mean necessarily that it is not possible to analyse the past by drawing on current scientific terminology, but it does mean that we are explicit about our intentions.[8]

A second obstacle that must be heeded is the tendency to treat past forms of sexual ambiguity too benevolently. This kind of retrospective utopianism is in part a result of the relativization of the present that accompanies all historical analysis of sexuality. It is an error that Foucault appears to succumb to when he considers hermaphrodites in the Middle Ages and Renaissance whereby supposedly, on reaching adulthood, they can elect the sex they wish provided they stay faithful to that sex for their whole lives.[9] In fact, this practice, which was permitted by certain legal traditions, only applied to exceptional cases such as those hermaphrodites whose predominant sex on birth could not be identified. Laqueur, for his part, has detected this excessive utopianism in Foucault's work. Given that the distinctions between the genders were not founded upon any biological differences between the sexes, these distinctions, as in any patriarchal order, were based on strict prohibitions and punishments that would be applied to any transgressor.[10] In other words, the extensive understanding of hermaphrodites as naturally occurring possibilities and not monsters in human form did not in reality allow greater flexibility or rights for these subjects during the modern period in comparison to their treatment from the nineteenth century onwards. In this century, they became teratological figures whose deformities hid their 'true sex'.

Thirdly, it is important that historical research does not fall into another trap in the understanding of this phenomenon. On the one hand, the attempt to discover a hegemonic model, a monist schema or 'one-sex' model, a dualist interpretation or the visualizing of three sexes at any given time, should be avoided.[11] On the other hand, it would be a mistake to avoid trying to perceive certain idealized types to aid analysis on the basis that the enormous diversity of models presented by the sources would impede such an undertaking. The first option could lead to a certain degree of dogmatism whereby the documents are only read in accordance with a priori understandings in an ad hoc manner and any evidence to the contrary safely eliminated. To some degree Laqueur has followed this route, as have some of his detractors.[12]

The second scenario would result in a kind of diffuse empiricism whereby models were rejected, resulting in a straightforward reading of what is said in the documents. Such a stance can be seen at times in the work of Joan Cadden in her examination of the plurality of understandings current in medieval medicine.[13] Nevertheless, it is true to say, as Cadden points out, that not all references to Aristotle's theories of generation in medieval times imply the hegemony of a dualist model of the sexes, which was opposed to the recognition of sex changes or real hermaphrodites.[14] But it is important to distinguish between the purely empirical reality of quotations from Aristotelian works and the ideal type that authors such as Lorraine Daston and Katharine Park designate, as a form of abbreviation (in the same way as Max Weber wrote about the 'protestant ethic'), the 'Aristotelian model' whose roots can be traced in *De Generatione Animalium*.[15]

The task, therefore, is not so much to identify the model of sexual representation that satisfies all epochs, or to reject any such possible model by abstract recourse to the intractable complexity of human creativity, capable of thinking anything at any one moment. The construction of models is not the end but the starting point of analysis. It is only by means of a critical comparative analysis, bearing in mind changing geographical and cultural realities over time, that models can be assessed for their validity and usefulness in historical analysis.

In this sense, we should be careful when analysing the culture of the *ancien régime* not to fall into an anachronistic interpretation of sex and gender. Following Laqueur, we know that in Europe in the sixteenth and seventeenth centuries strict differences between the social sexes, that is, in terms of gender, were not based on biological differences.[16] To be a woman or a man was not so much to possess a particular biological quality but rather to display a social attribute. In sources recounting news of prodigies or tragic occurrences (the 'relaciones de sucesos') in Spain in the sixteenth and seventeenth centuries it is fairly common to find items on sex change and the existence of hermaphrodites. In this kind of literature, sex is referred to as a 'habit' or 'state' (the famous Elena de Céspedes around 1587 speaks of 'taking the habit of a man'; Catalina de Erauso, who decided to live as a man in 1600 speaks of 'declaring her state', and the nun Fernanda Fernández in 1792 says 'I have taken the habit of a man' in order to refer to her change of sex).[17] To be one sex or the other was like belonging to a particular rank or social status. To change sex was to take a different kind of state, similar to the transition between singleness and marriage. In a similar way to being a noble or a vassal, one was a man or a woman. To belong to one or the other meant a series of privileges or prerogatives that the other did not have. Just as someone could not carry a sword or display certain signs of privilege, in accordance with rules governing clothing and presentation,[18] neither could a man dress as a woman or vice versa, except in exceptional circumstances such as in the theatre, masquerades or as a result of special permission granted by an ecclesiastical authority. But in the light of this, it cannot be said that in the *ancien régime* sex was subordinated to gender or that society somehow subscribed, *avant la lettre*, to a 'social constructionist' perspective.[19] Instead, the distinction between sex and gender was meaningless.[20]

First of all, the biological was never presented as the purely biological or as 'bare life'. The biological was understood from two angles: on the one hand it expressed a transcendant order, that of Nature as a moral sphere ruled by God. The occurrence of extraordinary or 'marvellous' events expressed, as part of a tradition that went back to St Augustine, the omnipotence of divine will.[21] The same occurred with the birth of hermaphrodites or with sex changes (understood as 'improvements') from women to men. At the same time, divine Creation was thought to be manifest in human reproduction and this demanded the existence of fully differentiated men and women. There was no contradiction between a form of biology that allowed for intermediate figures in terms of sex and an institutional framework that excluded them in gender terms. Both possibilities were inscribed in Nature, which was understood as a manifestation of divine will. For this reason, it has been argued that 'biopolitics', a power that immunizes 'bare life', can only come about in the gap left by the disappearance of the previous theologically based order.[22]

On the other hand, the second understanding of the biological, in addition to this vertical conception, linked the body and personal identity by means of a horizontal network of lineages, corporations and familiar relations. One's name, rights, obligations and prerogatives involved the body in a network of honours and dependencies.

This 'deployment of alliances'[23] implied a particular regime of visibility. As such, faced with the physical presence of an unfamiliar person, it was not a question of deciphering their true self or authentic person but rather of discerning from which family or house they came. It was a case of identifying the signs that showed their rank and if they were allowed to carry these signs *de iure*.[24] This was in order to prevent fraud and cases of false identity, something that was liable to affect courtly, community and family relationships.[25] For this reason, in the case of hermaphrodite persons or changes of sex by the authorities (in the form of midwives, doctors, judges and bishops) prouncements did not rely on any supposed deep-down real sex but on the confirmation of rank, dress and the occupation that the individual could legitimately assume.[26] In this sense, the physical aspects of a particular body functioned as a sign of rank and not as merely biological attributes. The act, inscribed in the tradition of Roman law, of assigning the 'predominant sex' rather than the 'true sex' of the newborn hermaphrodite individual obeyed this kind of concern: to determine the rank of the person, their prerogatives and associated obligations.[27]

The body, therefore, at a time when there was still no obvious schism between popular and elite culture, was not understood as a biological reality *tout court* or as a separate sphere dividing the self from the rest of the world. As the 'grotesque body' explored by Bakhtin indicates, this was an exteriorized reality, a microcosmos linked to a macrocosmos through lines of influence and preferences and dislikes.[28] It functioned as a text where divine design or the honour of lineage could be read.

The fluidity and changeability of this body whose sex could be transformed as a consequence of an abrupt shift in the person's activity or work was such that change of sex, bearded women and menstruating or lactating men could all occur. They occurred as an expression of divine will whose omnipotent power was capable of moulding the individual accordingly. In this sense, these cases were 'marvels' (*mirabilia*), strange occurrences that certainly deviated from the normal order of Nature (*praeternaturalia*) but which could not be understood as being against nature (*contra natura*). Indeed, as we will see, a significant section of Spanish thought in the sixteenth and seventeenth centuries understood these intermediate beings not as monstrosities but as possibilities perfectly within the broad dynamics of a prolific and diverse expression of Nature.

However, any ambiguity and mobility between the sexes was subject at the same time to severe restriction. God had wanted two sexes to exist in humans in order to guarantee procreation. The task of the civil and religious authorities was to watch over the border between the sexes and to set out the criteria whereby any one intermediary individual could participate in both sexes. These same authorities would also attempt to dissuade and to punish those who attempted to transgress those established limits by making ambiguity a *modus vivendi*. For this reason, it was not uncommon that at the time of the Counter-Reformation hermaphroditism and sex change came to be associated with sodomy as the counter-natural activity par excellence.[29] In these cases, what are invoked are not the marvels that exalt divine power, but the maleficience (*magicus*) inherent to these events, signs of sin or a warning of danger to come.

But any more or less natural or praeternatural instance of *mirabilis* and counter-natural *magicus* is only part of what the ambiguous body could represent. A third type of experience that was doubtless less frequent nevertheless draws on the same idea of metamorphosis and sexual ambivalence. This was the *miraculus*, a supernatural intervention that departed from the normal course of Nature and presented a salvationist message. The *miraculosus* could include saints who changed sex as a result of divine intervention, the final resurrection of women converted into men and, on a different level, the invocation of the androgynous as a symbol of perfection uniting two contrary tendencies.[30]

Mirabilia

The category *mirabilia* designated extraordinary beings and events, 'marvels' that showed the omnipotency and the inscrutability of divine design. This tradition stretches back to the Augustinian text *De Civitate Dei*, and was followed by St Isidore's *Etymologies*.[31] Portents are not necessarily counter-natural beings or isolated cases with no significance; they are natural rarities that always have their analogies in the Universe. The Universe is conceived as a dense network of relations which reveal a hidden harmony, known by God but of which humans are ignorant. Because of this ignorance, humans perceive these figures as disconcerting and horrific.

It has been argued, only partly successfully, that in the later medieval period, as a result of the social and cultural crisis that took place in the fourteenth century in the wake of plagues, massacres and famines, the belief in the harmony of the Universe was undermined. The presence of the monster was understood as evidence of the work of the devil and of great calamity to come.[32] It has also been argued that this understanding would give way, in the light of the coming of science, to a naturalist perception of the monster in the context of an emerging literature on 'marvels'. This literature stimulated devotion to pleasure and a certain curiosity about this figure, which was presented as evidence of the benevolence of God working through a prolific and diverse expression of the natural world.[33] This image would constitute the preface to seeing the monster as an error of nature particularly from the seventeenth century onwards. This error of nature would be a deformity that could be explained by purely immanent causes. In this way, a process of disenchantment and rationality would be complete.[34]

But things are not that simple. Throughout the Middle Ages there prevailed a certain division between the representation of the monster as a species from exotic lands and the idea of the monster as a warning of disaster. This dual conception would come, on the one hand, from travel literature, centred on the description of marvels (species that showed the unlimited power of the Creator) and, on the other hand, a literature of prodigies that presented monstrous individuals, not species, as a sign of evil.[35] From the sixteenth century this distinction became less clear. Both portentous individuals and species could be understood to show the hidden harmonies of the cosmos thus proving divine will. At the same time, the tradition that saw the monster as a punishment from Providence or a sign of disaster to come survived.[36]

Rather than this teleological account, which argues for a sceptical and disenchanting modernity, the sixteenth and seventeenth centuries present a number of mixed understandings that exalt the monster as a marvel and at the same time fear him as a manifestation of evil. This ambivalence is clear in the case of the hermaphrodite and changes of sex, despite the fact that most medical opinions classify these as natural happenings. They are still understood to represent something out of the ordinary or rare and they are represented as such in the Spanish case in the literature on marvels and in the 'relaciones de sucesos'.

The genre of the literature on marvels maintains its presence throughout the whole of the Spanish modern period,[37] and even has its expressions beyond the end of the seventeenth century.[38] Examples of this kind of literature, mainly penned by ecclesiastics during the period 1540 to 1677, clearly show the interest in sexually intermediate persons described as 'marvels'. The notion of *mirabilia* is maintained throughout this period in this kind of text. The apparently deformed and disordered show the limitations of human intellect and the complexity and brilliance of the order imposed by God, however inscrutable to human eyes.[39]

Hermaphrodites and episodes where masculinization occurs (in contradistinction to 'feminizing' incidents) are presented in these sources under the rubric of 'natural' events, even though they are considered to be extraordinary. Usually, individual cases are referred to, although there are also, following the medieval tradition of travel literature, accounts of entire exotic peoples who have the reputation of hermaphroditism. The work of Pliny, as presented by the philosopher Calliphanes, falls into this category.[40]

The description 'natural' as used to refer to hermaphrodites, manly women or *viragines*[41] and masculinized females[42] is not the same as our description of such phenomena as 'biological'. In this literature sex is identified as a variant of rank. The best example of this is offered by Antonio de Torquemada (whose work is discussed by Martín del Río and Juan de la Cerda).[43] Torquemada referred to the case of a woman from Condado de Benavente (Zamora), who was married to a poor labourer. One night the woman decided to leave her husband, disguised by some clothes stolen from a local boy. She adopted the lifestyle and work of a man and the force of these new habits was such that 'ella se convirtió en varón, y se casó con otra mujer' (she changed into a man and married another woman). This woman, 'hecha varón' (having been made into a man), was spotted by an acquaintance who was asked to keep her secret.[44]

The change in clothing and occupation is what unleashes her sexual transformation, either through the action of imagination or that of nature.[45] It is as if her abandonment of her state as a married woman and the taking on of the appearance of a man brought about her bodily transformation; it is as if the physical were an external expression of her 'state'. In addition, in the light of the uncertainty that surrounded the possession of one state or another in terms of right, duties and obligations, her act of fraud consists in taking on privileges that do not correspond to her. But her status is not reducible to either biological sex or gender. Gender does not provide the foundation of her sex because her 'nature' is only distinguished on the basis of her dress or occupation. It is as if sex and gender were undifferentiated, making up something else: her rank in society, something stable in itself in an ordered society but susceptible to disordering and confusion in exceptional cases.

When discussing the natural character of hermaphrodites and virilized women all these texts follow similar formats. First of all, they present their structure of argument, like the *quaestiones* or as in examinations of conscience (Martín del Río, Antonio de Fuentelapeña), they opt for a dialogical form (Alonso de Fuentes, Antonio de Torquemada, Juan de Pineda) or they adopt a narrative description (Pedro Mexía, Juan Eusebio Nieremberg, Juan de la Cerda). They all cite ancient authorities on these matters (for example, Pliny, Ovid, Hippocrates, Phlegon, Aulus Gelius and Livy) and modern authorities (Joviano Pontano, Amato Lusitano, Fulgoso, Montaigne) who affirmed the existence of hermaphrodites and masculinized women.

In addition, these authors display an eclectic array of shared medical knowledge. This includes the Hippocratic theory of humours and generation, references to Galen and Avicenna whose accounts presented the male and female organs as identical in structure but not in position,[46] and the mention of the Aristotelian teleological principle which accounted for masculinization of women as 'improvements'. In the Aristotelian vein too was the discussion over whether the woman was a weaker form of man or not.[47]

In this mixture of positions, the Hippocratic one-sex model prevails, although accounts are much nuanced. The two accounts of 'marvels' that display most medical arguments, that of Fuentelapeña and especially that of Martín del Río, seem to have come under the influence of the French medical doctor André du Laurens (1550–1609). Du Laurens was the author of *Historia Anatomica Humani Corporis* (1593) and defended, according to Michael Stolberg,[48] the Aristotelian model of the two dichotomous sexes. The reading of this work, clearly first-hand in the case of Martín del Río in 1606 and perhaps second-hand by Fuentelapeña, sounds a cautionary note over the supposed hegemony of the one-sex model in the Spanish case.

In the case of Fuentelapeña, who, in contrast to strict Aristotelian thought, nowhere doubts the existence of true hermaphrodites, the explanation of this 'marvel'[49] draws on a clearly Hippocratic account although it is taken from Albertus Magnus.[50] The allusion to André du Laurens ('Andrés Lorenço') is occasional and serves merely to ratify his own thesis. This is that women who change into men are in reality 'hidden hermaphrodites', beings that possess two natures although this has become visible only because of excessive natural heat.[51] Strictly, then, sexual transformation is not possible. Fuentelapeña includes Du Laurens (and Martín del Río) amongst those who hold this opinion. He is aware that Du Laurens emphasizes sexual difference thus placing himself against the majority which was in favour of the Hippocratic-Galenic paradigm of the single sex. Fuentelapeña finally, however, seems to move towards an Arabic-Galenic alternative which, without admitting true sex changes (which are understood as manifestations of the invisible sex of hidden hermaphrodites), supports the 'one-sex' model:

> que aunque el instrumento sea único, puede invertirse de adentro afuera, como un guante ... las mujeres tienen los mesmos vasos seminales y órganos que sirven a la generación, que los hombres.[52]
>
> although the instrument is one, it can be inverted from inside towards the outside, like a glove ... women have the same seminal ducts and organs of generation as men.

The case of Martín del Río is different. In the second edition of his work *Disquisiciones Mágicas*, from 1612, he declares that he has read the *Historia anatómica* (1606) by Andrés de Lorenzo.[53] This reading, Del Río avers, confirmed his own understanding that supposed masculinized women were in fact 'hermaphrodites' who possessed both sexes. As a result, the transformation of sex would be nothing more than the exteriorization of the male nature that had been hitherto hidden.[54]

However, the work by Du Laurens goes further than affirming the existence of hermaphrodites. Martín del Río tells us that Du Laurens effectively rejects the one-sex paradigm and that the genitalia of one and the other sex are completely different 'no sólo por su situación, sino por su número, forma y estructura' (not just in their position but in number, form and structure).[55]

The renowned demonologist and Jesuit appears to subscribe to this dualist model, a move that allows him to reinterpret two elements associated with the one-sex model. The first of these is that the woman is a failed or weakened form of man. Secondly, he revises the teleological notion that nature always tends towards perfection.

With respect to the first matter, he suggests, following Du Laurens, that woman is not an incomplete man as in the hierarchical and vertical system of the one-sex model but a finished organism that possesses its own structure. This structure corresponds to the function of womankind in order to preserve the species.[56] On the question of teleology, he provides a different interpretation:

> *hay que decir, pues, que la naturaleza siempre procura lo más perfecto, no porque siempre tienda a engendrar varón, sino porque cuando tiende a ello procura hacerlo lo mejor posible y lo mismo cuando se propone hacer hembra.*

> we can say, then, that nature always seeks the most perfect, not because it always seeks to engender a male but because when it seeks the perfect it does so in the best way and in the same way as when it tries to make a woman.[57]

By means of a reading of the work of Du Laurens, Martín del Río, despite not adhering to an Aristotelian dualist model (he concedes the existence of hermaphrodites), does dissent from the Hippocratic-Galenic tradition by recognizing the irreducible singularity of the female body.

The same kind of register as that used in accounts of 'marvels' can be seen in a case of sex change that took place in 1617 in the Convento de la Coronada, in Úbeda.[58] It was mentioned in a *relación de sucesos* published the same year. As is usual in the majority of texts of this type it appears in the form of an epistle,[59] a letter sent from the prior of the order of Saint Domingo of Úbeda to the abbot of San Salvador, Granada. In contrast to other cases of prodigies and apparitions, in this case there was no intention of terrorizing the reader into seeking conversion or sanctuary in the Church.[60] Like other accounts in this genre of literature this text presents the masculinization of a nun. This is presented as something extraordinary but natural, to be included within the 'miracles of nature'.[61]

The nun in question was María Muñoz, who was suspected of being a man by her sisters. She was examined by the prioress who declared her to be a woman. However, shortly afterwards, the nun wrote to the prior of Saint Domingo of Úbeda. After interviewing her and summoning the Dominican prior from Baeza the authorities decided to examine her:

> *y hallamos ser hombre perfecto en la naturaleza de hombre y que no tenía de mujer sino un agujerillo como un piñón más arriba del lugar donde dicen que las mujeres tienen su sexo a pie del que le había salido de hombre.*

and we found a man perfect in the nature of man and there was nothing of a woman apart from a small hole like a pine-nut above the place where it is said women have their sex next to where a man's sex had come out.

The text shows a certain degree of familiarity with the medical knowledge of the period. The nun's sex had been changed as a result of some strenuous work in the fields; this had meant a sudden increase in heat and the expulsion of a penis from her body. The exertion provoked 'un gran dolor entre las dos ingles' (a great pain in the crotch) which had caused a swelling. After three days, the swelling went down but 'le había salido naturaleza de hombre' (a man's nature had come out). Seven days afterwards the transformation gathers pace: 'le comenzaba a negregear el bozo y se le mudó la voz muy gruesa' (the facial hair darkened and the voice became much deeper). Although the document is not entirely clear on this point the illustration that accompanies the case suggests that the nun was indeed a hermaphrodite.[62]

Be that as it may, the episode confirms that the text belongs to a period that did not distinguish between biological sex and gender. Such is confirmed by the attitude of the father of María Muñoz. His first reaction was of shock but this did not last: 'el padre está muy contento, porque es hombre rico y no tenía heredero' (the father is very happy as he is a rich man and had no heir). It is clear from such statements that sex was not understood as a deep seated biological reality. Instead, it was related to kinship, blood lines, worldly goods and names that made up a lineage. Nowadays, the difficulty for the parents of the transsexual is to accept the psychic peculiarity of their offspring so that their mental health and self-esteem are not harmed; in the seventeenth century the masculinized woman raised the question of whether there would need to be a change in the transmission of inheritance and the family name and if it affected the rights of other descendants.

María Muñoz herself received her new sex not as a kind of physical novelty but as the promotion to a higher rank from nun to first-born and only heir. This would afford her access to privileges previously thought impossible. This change in circumstance, in an order based on status that was not used to sudden alterations, merited being communicated to the highest temporal authority that could accord such privileges. For this reason the author of the letter recommends that the King should be written to in order to communicate the 'strangeness' of the case.

Galenisms and Aristotelianisms

Historians of Spanish medicine agree that throughout the sixteenth century Galenism was the primary medical doctrine.[63] But this was an unusual form of Galenism which was derived less from the great medieval commentaries, including Arabic texts such as Avicenna's *Canon*, than from a contrasting reading with the classics of Greek medicine.[64] The humanist return to the original texts had also been followed in the area of medicine. Galenism, enriched primarily by Hippocratic references,[65] and to a lesser degree through Aristotelian influences, dominated the medieval intellectual scene in medicine. In addition, the field was open to the anatomical teachings of Vesalio and other Italian masters.

This hegemony of the Galenic tradition undoubtedly led to the predominance of the one-sex model, a position that was practically uncontested throughout the second half of the sixteenth century. This does not mean, however, that Aristotle's thought on the sexes and generation was ignored in the Renaissance period in Spain. What happened was that Aristotelian understandings did not significantly undermine Galenic-Hippocratic predominance and its one-sex model.

Within this framework, women were described as having the same organs of generation as men. The difference resided in their position. They were internal to women but external to men. Another matter was that of temperament, which was related to the positioning of these organs. The lack of heat due to the coldness and the moistness of women explains why the organs such as the 'madre', analogous to the 'verga', and the 'testicles' or 'compañones', remained inside the body. This internal arrangement was in order to allow conception to take place. Doctors not only invoked Galen to prove this point but referred to observation and illustration. Anatomists such as Bernardino de Montaña (1489–1558),[66] Juan Valverde de Amusco (1525–88),[67] Juan Fragoso (1530–97)[68] and Andrés de León (1560–1602),[69] theologians who wrote texts on the 'conservation of health' such as Blas Álvarez de Miravall (*fl.* 1597),[70] or doctors who published influential books such as Juan Huarte de San Juan (1529–88),[71] contributed to the consolidation of the representation of woman as an imperfect man.

The recourse to the idea of one sex corresponds to an understanding of generation that combines Hippocratic and Galenic theory. The former insisted that generation was the result of the mixing of male semen (predominantly warm and dry) with female semen (moist and cool) in the interior of the 'mother'.[72] This is usually represented as formed by seven cavities. If during the mixing process the male semen was dominant and contact occurred in the right three cavities a male would be born. For this reason mothers were advised to lie on their right sides if they wished for a male child. If the female seed was victorious and the fusion took place in the left three 'cells', a girl would be born.[73] If during the process neither the male nor the female type of semen prevailed and the mixture occurred in the central cavity a hermaphrodite would be born. Hippocratic theory required the emission of female seed for fertilization to take place. It was for this reason that doctors following this doctrine referred to the lack of sexual appetite among sterile women.[74]

On the other hand, the Galenic doctrine, which was adopted by an important group of Spanish doctors, argued for some differences with respect to the Hippocratic model. First, the more important role and the superior strength of male semen in terms of its ability to fertilize over that of female were emphasized.[75] It was also understood that the woman, in addition to living seed, supplied menstrual blood which played a crucial role in the formation of the foetus's body.[76] This understanding of female sperm as seed, although of inferior quality, distinguished Galenic theory from that of Aristotle, which favoured the male seed on its own as having the ability to fertilize. Female seed was reduced to mere primary material.[77]

Within this framework faithful to the one-sex model, which dominated Spanish thought in the 1500s, a huge range of intermediates were understood to exist. Between the most perfect male (the virile man) and the most imperfect (the feminine woman) the existence of hybrid and itinerant figures was permitted. These included hermaphrodites or androgynes, whose existence was supported by a long philosophical and literary tradition and whose formation was explained by recourse to the Hippocratic mixing of semen. Not only was their existence accepted and believed to be common in exotic lands;[78] they were classified into several categories.[79] Indeed, their presence in Castile was also recorded.[80]

The prestigious Leonese anatomist, Luis Mercado (1525–1611), who became court doctor under Felipe II, appeared to dissent from the majority current by considering hermaphrodites in the third book of his *De Mulierum Affectionibus* (1579) amongst the varieties of monster.[81] Following the arguments of *De Generatione Animalium*, Mercado explained the monster as a result of a disproportion between material and form (*vis formativa*) which had taken place in the generation process. The result was the disruption of the similarities between offspring and parents.[82] However, reinterpreting Aristotle on this point, Mercado did not consider the monster as a being contrary to the intentions of Nature. The monster was a 'praeternatural' figure, that is, it only deviated from the order established by Providence to the extent of what was usual in that order.[83] Nowhere does he deny the existence of true hermaphrodites, although he believes that any claims of sex change are fables. These would come about from the confusion arising from disproportionately large labia in some women or from cases of a prolapsed uterus. Finally, the beard of some women was nothing more than an indication of the suppression of menstruation.[84]

Together with the hermaphrodite the existence of other transitory forms within the one-sex model was recognized. These included menstruating males,[85] manly women or *viragines*,[86] 'soft' men ('mariosos')[87] and lactating men.[88] Finally, there was frequent mention of individuals who had changed sex. These were nearly always cases of masculinization; not all doctors excluded the possibility of feminization, although this was considered exceptional.[89] In the same way as in the literature on marvels, past and contemporary accounts were cited that gave credence to these stories. The similarities and symmetries between male and female organs and recourse to the humoral theory in order to explain alterations in the humours, as a result of an excess or lack of heat, supplied a coherent basis from which to explain cases of sexual metamorphosis.

From a close reading of Spanish medical texts from the seventeenth century we can see that some of these postulates on the representation of the sexes began to change. While we cannot talk of an intellectual break as such, as the anatomical textbooks of the previous century continued to be republished and the Galenic dominance continued practically undisturbed, we can point to some developments at the beginning of the century. These concerned the differences between the sexes and they affected understandings of cases of hermaphroditism and transmutation of sex.

It seems certain that the work of André du Laurens (1558–1609), *Historia Anatomica Humani Corporis*, published for the first time in 1593, played an important role in this development. Du Laurens, of

Jewish descent, professor at Montpellier and physician of Henri IV, questioned the identical content and structure of the male and female body in Galenic theory and denied the possibility of sex change. This does not mean that he subscribed in any simple manner to a two-sex dichotomous model in accordance with the Aristotelianism of Jean Riolan,[90] as Stolberg has supposed.[91] But nor does it mean that his differences with prevailing Galenic-Hippocratic hypotheses can be minimized, as Laqueur would have it.[92]

There were a number of doctors who agreed with the arguments of Du Laurens. These include Pedro García Carrero (1555–1628), professor at Alcalá de Henares, and court physician of Felipe III. García Carrero is considered by historians as 'la personalidad más importante de la escuela médica complutense a comienzos del siglo XVII' (the most important figure of the Complutense medical school at the beginning of the seventeenth century).[93] This author examined the question of generation, the anatomy of the sexes and the genesis of monsters in his extensive *Disputationes Medicae* (1605). As we noted in the case of Luis Mercado, the scholastic influence and Aristotelianism were present in a work whose argumentative structure followed the university convention of the *quaestio disputata*.

García Carrero discussed sexual difference but energetically denied that woman could be considered as a failure of Nature or as an imperfect male, a sort of monster.[94] He understands female anatomy to be composed with the aim of generation. The female body is not a mistake or an error, although he does admit that the female constitution is weaker and colder than that of the man. Further, he recognizes, along with Aristotle, that the female body only fulfils the role of material cause in generation.[95] García Carrero rejects, therefore, one of the elements of the monist model – the understanding of women as an imperfect male.

In addition, García Carrero disagrees with Galen in that the difference between men and women goes beyond the mere position of the genitalia (internal in the woman, external in the male). He suggests that 'differunt enim aliter', although he does not specify precisely what these differences are. He also dedicates a number of chapters to the question of the monster and the genesis of different varieties. Here, García Carrero adheres fundamentally to the argument sustained by Aristotle in his *De Generatione Animalium*. But he differs from Aristotle in his consideration of the monster not as an error of nature ('deffectus naturae'),[96] since Nature can never be wrong, but as an entity *praeter naturam*, that is, a deviation with respect to the normal order provided by Nature. He relies on Aristotelian theory in order to explain the monster as a consequence of disequilibrium (of magnitude, quantity, etc.) between the *vis formativa* and the material used for generation. From this perspective, the hermaphrodite is classified as a monster; a creature with both sexes which can appear in human and animal species.

Following the argument of *De Generatione Animalium*, hermaphrodites would be the consequence of an overabundance of material supplied by female menstrual blood. This excess would not allow for the formation of two distinct foetuses although more material than required for one foetus would be present. This material, where the male semen was predominant in certain parts and the female semen in others, would go to form the genitalia of the other sex. In any case, according to Aristotle, the hermaphrodite would only be apparent since the sex of the foetus did not depend on the genitalia but on the heat of its heart.[97]

García Carrero includes the hermaphrodite amongst monsters and, even if he is not specific on their genesis, given his Aristotelian references, it can be argued that the Spanish doctor believed that excess material was indeed the principal cause.[98] On the other hand, he does not say explicitly that true hermaphrodites are a fiction and he does not allude to the determination of sex by means of 'cordial' heat. However, for García Carrero, in hermaphrodites one sex is present in atrophied form and the other active.[99] Does this mean that García Carrero denies the existence of human hermaphrodites? Rather than this, it would appear that García Carrero, under the influence of Aristotle and Du Laurens, whom he cites frequently,[100] embraces a perspective that does not tie in exactly with the one-sex model prevalent during the preceding century.

The influential juridical figure Alonso Carranza (*fl.* 1625) trod a similar path to García Carrero. Carranza was the author of *Disputatio de Vera Humanu Partus Naturalis et Legitimi Designationi* (1628),[101] and he argued that hermaphroditism was a monstrous condition. Citing André du Laurens amongst others, he considered hermaphroditism to be 'peccatum in sexu',[102] that is, an 'error' of sex.[103] However, in contrast to those who considered hermaphrodites to be a mere *figmenta*, or fiction or simulacrum, Carranza believed in their existence in reality.[104] He classified them according to four types:

masculine, with a fully formed male sex but with a false vulva that distilled no liquid whatsoever; feminine, with a vulva and menstrual flow, who possessed above their vulva a sort of male member which underwent a species of erection but without testicles or scrotum and which was incapable of producing semen. The third type possessed both sexes located in opposite places and even though they emitted a kind of semen and were capable of miction could not procreate. The fourth class consisted in those that had perfectly formed male and female genitalia and were 'potent'. They possessed on the right side a male breast and on the left a woman's breast.[105]

A further step in the critique of the one-sex model was taken by Spanish medicine during the 1600s by Gaspar Bravo de Sobremonte (1603–83). He was professor at the University of Valladolid, physician in the court of Felipe IV and Carlos II and primary medical adviser for the Inquisition. He was one of the most important anatomists in Spanish medicine of the time.[106] In a 'promptuario' in the third volume of his *Opera Medicinalia*, published in 1671 and in one of the three *disputationes* that make up the fourth volume, published in 1679, Bravo de Sobremonte discussed the issue of hermaphroditism and sex change.[107]

After surveying the different authorities on the question and a number of accounts that alleged the existence of sexual transformation (Ovid, Plato, Herodotus, Hippocrates, Paré, Montaigne, Torreblanca Villalpando, Nieremberg, Martín del Río, etc.) and after referring to several Spanish cases (the famous Brígida de Peñaranda and metamorphoses that occurred in Madrid, Alcalá de Henares and Córdoba), Bravo de Sobremonte outlined his arguments against the reality of such events as impossible in Nature. In order to argue this point he began by refuting the Aristotelian *dictum* as already seen in the work of García Carrero: Nature does not produce woman through error as though a mistake had been made while trying to create something more perfect, that is, the male.[108] The female is not a monster or a failed male but instead a figure engendered by Nature who possesses certain peculiarities allowing her to procreate and thus guarantee human survival.[109] For this reason, the female was not essentially different from the male, she was no monster. Any difference was merely accidental.[110]

In arguing for this difference, Bravo de Sobremonte attacks Galenic thought on the subject: 'Scrotum ut utero non solum differunt in situ, sed etiam in conformatione partium, numero, magnitudine, et insitis viribus: ergo non sunt una entitas, ut supponit Galenus.'[111] No refutation could be clearer – the distinction between male and female genitalia makes sex changes impossible because any excess of heat in the case of the expulsion of the female organs to the exterior would not be sufficient to transform them into male organs. The difference is not just one of position; it affects the attributes of both sexes in many senses. Bravo de Sobremonte does not merely argue for this duality but carefully compares the male and female organs (penis, uterus, testicles and scrotum) in order to back up his thesis of difference.[112] Amongst the authorities to which he refers are García Carrero, André du Laurens and Jean Riolan.

Any sex change or conversion of the organs of generation from one sex to the other is impossible given their incommensurable differences.[113] How are women who apparently change into men explained? Bravo de Sobremonte suggests three explanations. First, these women could be men who look like women and whose genitalia have not descended (what might be termed now cryptorchid males). When the genitalia do in fact descend these men appear to change sex even though they have always been men. Secondly, there may be cases of women whose labia or clitoris (Bravo de Sobremonte does make the distinction) are so well developed that they are taken to be men. Thirdly, any confusion could arise from the presence of 'hermaphroditae seu androgines' – both terms are used – who can move between the two sexes and are identified wrongly as having changed sex.[114]

Those individuals who do change sex belong to the realm of 'poetarum figmenta', literary fables or figments.[115] However, Bravo de Sobremonte does not deny the possibility of real hermaphrodites and even cites these cases as a source of error in supposed cases of changed sex. Bravo, who so firmly argued for the dichotomous duality of the sexes, does not go as far as Neo-Aristotelians such as Constantino Varolio and Jean Riolan,[116] who reject completely the existence of true hermaphrodites. In the second part of the 'Promptuarium XXIV' of 1671, whose precise title is 'De hermaphroditis', Bravo de Sobremonte clearly admits that such individuals exist. Although he recognizes that the medical authorities usually include them in the monster category he indicates some types of hermaphrodite which have a particularly monstrous conformation.[117] He adopts the classification of 'androgini' proposed by

Alonso Carranza and suggests four different types. He rejects the astrological explanation of the hermaphrodite and favours the 'imaginative' cause during coitus. He also has recourse to supposed imaginary cavities in the uterus which were present in the Hippocratic tradition. Despite this variety of causes, he seems to go for a Hippocratic explanation – the conformation of a hermaphrodite depends on the mixing of the seminal material and on the balance of forces between the male and female principles. Finally, confirming the impossibility of any masculinizing metamorphosis, Bravo insists that hermaphrodites possess the two sexes *ab initio* and that even they do not undergo any kind of transformation. What actually happens is that the male genitalia remain hidden until the individual develops and they become visible along with the female parts.[118]

Any critique of the one-sex model such as those contained in the work of García Carrero and Bravo de Sobremonte should not be taken as signs of an emerging rationalism that would finally exile fantastic figures such as the hermaphrodite from western thought. On the contrary, these figures would continue in good health in the medical literature of the 1600s.[119] Even astrological, moral or theological explanations for the monster would continue and would inform not only miscellaneous texts on 'marvels' but medical thought too.

An example of this is offered by the work of José Rivilla Bonet (*fl.* 1690), *Desvíos de la naturaleza o tratado del origen de los monstruos* (1695).[120] Rivilla was surgeon to the Viceroy of Perú and at the Royal Women's Hospital de la Caridad de Lima, which published this work. It was here that the author had had the opportunity of observing in November 1694 the birth of a bicephalous monster. This episode not only challenged anatomical knowledge to date but also posed a theological dilemma as to whether the new-born in question should be baptized once or twice. It was this conundrum that led Rivilla to write his book.

The book is divided into ten chapters.[121] The two first chapters discussed the significance and definition of the word 'monster' and its relation to a family of words that were semantically close, such as *portentum*, *ostentum* and *prodigium*. The following two chapters sketched out the taxonomy of monsters. Chapters 5 and 6 analysed the causes of such deviations. The seventh chapter looked specifically at the formation of bicephalous monsters. The final chapters constituted a detailed study of the monster born in Lima. Its anatomy is described and the question of how many souls should be ascribed to the monster is discussed and whether baptism should be performed on both heads. The book finishes with an appendix in which Rivilla Bonet gives an overview of different cases he has come across, together with the array of surgical methods employed in each.

This text is situated on the borders of the medical treatise and the narration of curiosities or 'marvels'. In the tradition of other authors such as Riolan, Du Laurens and Bahuin, who are cited, Rivilla rejects the Aristotelian definition of the monster. The monster should not be categorized as an error but as a being 'procreado fuera de la intención de la Naturaleza'.[122] To the degree that this being is a result of the process of generation, a process ordered by Nature, it can be considered 'contra la naturaleza, mas no contra toda ella, sino contra su más frecuente caso' (against nature but not against all of her, but against her most frequent expression).[123] The Lima birth was a case of a child 'praeter naturam' and not 'contra naturam'.

Despite discussing extensively the problem of monsters, Rivilla says little about sex changes and hermaphrodites. The hermaphrodite is mentioned on one occasion and this is just to remind the reader that animals that, given the characteristics of their species, present certain peculiarities such as 'entrambas naturalezas; ya hermafrodita con alternativa amba en los sexos' (between both natures; a hermaphrodite with both alternatives in the sexes) should not be considered monsters.[124]

When discussing the genesis of monsters Rivilla distinguishes between two types of causes. First, Rivilla discusses superhuman or diabolical causes. Here he connects with an age-old tradition that includes St Isidore and St Augustine. The monster is understood as a punishment for the abominable sin of the parents or as the forewarning of a calamity to come.[125] These causes also include monsters engendered through carnal intercourse with the devil or as a result of certain astrological alignments.[126]

The second type of causes covers 'inferior physical causes'. These are the main subject of the book. Rivilla is faithful to the explanation of monsters provided by the Aristotelian tradition, which is detailed, as we have seen, in the work by Luis Mercado, *De Mulierum Affectionibus*. The distinction between superhuman causes and physical causes does not in itself imply that both sets of causes worked independently. Rivilla remarks that physical causes can operate as 'secondary causes' which could also serve God for purposes of punishment.[127] In this way, Rivilla goes beyond the reasoning of monsters as divine artefacts or praeternatural rarities. The monster is not just an object of curiosity or repugnance but also one of terror because of its diabolical implications or its significance as a punishment for abominable acts. In this way, the monster seems to depart from the field of *mirabilia* to become part of the world of evil or *magicus*. This second category will be examined in the next section. For now, a brief summary of what has been said is provided.

Two different types of evidence where the problem of the configuration of the sexes occurs have been discussed. The first type of evidence is one of fuzzy borders and miscellaneous in nature and is known as the literature of marvels. This first genre tends to accept the existence of intermediate forms such as the hermaphrodite as a natural occurrence and as a sign of divine omnipotence. The Hippocratic-Galenic schema of one sex and the Aristotelian principle that understands natural dynamism as teleologically driven towards perfection are both present in this literature. This *corpus* tends to reject the possibility of women changing into men but accepts unproblematically the reverse, understanding such metamorphoses as 'improvements'.

At the same time, the medical literature, in a more explicit and systematic manner, in general subscribes to the same monist model that depicts women as imperfect males. It understands that female genitalia are similar to those of the male, following Galen, although they are positioned differently. In the light of this explanation, a whole range of intermediary figures is accepted (menstruating and lactating men, *viragines*, hermaphrodites and masculinized women). In general, both the literature on 'marvels' and medical treatises represent these persons as creatures that do not contravene the natural order. They can be portents that provoke admiration or curiosity, praeternatural rarities, or monsters (Mercado) that entail disgust and repugnance. Finally, in both types of text, sex is thought of as rank. It is not a pure biological fact or a social construct alone. Nature, subject to Renaissance *divinatio* as an expression of divine will, can at the same time produce marvels that alter the distinction between male and female and base the continuity of masculinity and femininity on the fact of human procreation produced by the difference between man and woman since the time of Adam.[128]

The general thrust of these ideas remains strong until the end of the sixteenth century. From the beginning of the seventeenth century, however, a number of question marks start to appear. From the field of demonology and 'marvels', the texts by Martín del Río and, to a lesser degree, by Antonio de Fuentelapeña begin to sketch out a challenge to the one-sex model and this is picked up upon in medical thought. Two major medical authorities in Spain in the seventeenth century, Pedro García Carrero and Gaspar Bravo de Sobremonte, adduce such interpretations. The main source cited in both works that provides the source of this change is that of André du Laurens, whose *Historia Anatomica Humani Corporis*, originally published in 1593 and known of in Spain from the early 1600s, questions the one-sex paradigm.

Something, then, is starting to change during the first decade of the century. But we should not be thinking in terms of a break with the old Hippocratic-Galenic one-sex model, as Stolberg argues. The most influential medical treatises of the previous century continue to be published and still enjoy a huge degree of acceptance in medical schools. Furthermore, although the idea that woman is an imperfect man is critiqued this does not mean by itself that social differences are understood to have a biological basis as in the medico-legal texts of the late eighteenth century. The framework of sex as rank remains unassailed as does the idea of an onto-theological Nature that is inserted into a transcendent order that provides it with meaning and moral finality. In this sense, we can talk of a kind of 'sexual *ancien régime*', which would exist in parallel to other expressions of the old regime in terms of demography, politics and economics.

So, where precisely do the intellectual breaks in the 1600s occur? First, the notion of the woman as a failed male or monster is placed in doubt. Instead, women are understood as being of the same species as the man with accidental differences but produced deliberately by Nature, who has given her a number of specific qualities that allow for the human species to continue. Secondly, such a notion entailed the rejection of the Galenic principle of sameness between male and female genitalia. Now, difference was emphasized, not only in terms of position but also magnitude, structure and number. Certain parts were signalled as specific to one sex, such as the clitoris, which have no correspondence with the male parts.

Thirdly, this argument eliminates the possibility of sex changes (see Bravo de Sobremonte and to some degree Martín del Río and Fuentelapeña). It also entails doubt with respect to the existence of true hermaphrodites (for example, García Carrero accepts the existence of hermaphrodites but argues that only one sex actually functions in them while the other is atrophied). These shifts do not allow for a two-sex model of incommensurable sexes and fundamental differences between men and women. Rather, they suggest certain cracks or anomalies within the one-sex model. These should not be exaggerated (Stolberg), but nor should they be minimized (Laqueur).

These anomalies will be resignified and reinterpreted in the light of the new understanding of sexual difference that is slowly formed from the eighteenth century onwards.[129] These shifts will become the demonstrative proofs of the new model. But for this to take place, Life and Nature will have to free themselves from the old order. *Divinatio* must become immanent causation and at the same time Life and Nature must become autonomous processes governed by their own internal logics. Only with the appearance of 'bare life' as the sphere in which social regulation occurs, where purely bological accounts allow for emerging political orders, will biological sex become the fundamental basis of sexual identity.

Magicus, Miraculosus

Magicus corresponds to what is supernaturally evil, where sin is present and where the intervention of Satan can be found. In this sense, sex changes and hermaphrodites often appear from the times of primitive Christianity up to the Renaissance as portents or warnings of some catastrophe to come or as the result of either sins against nature or sodomy. The hermaphrodite is understood as the creator of disorder and chaos that undermines the regulatory logic of Nature.

This state of affairs is shared by the hermaphrodite, sex change and the monster. The etymology of *monstruum* (*monere*) or of *portentum* (*portendere*) shows how this significance came about. Its evil aura comes about through the production of disorder or apparent rebellion against the harmony of nature set out by God. Both characteristics have a long history, which goes back to pagan times: the tradition of the hermaphrodite as a bad sign (Cicero, Pliny, Livy) and as an error of Nature (Aristotle, Lucretius, Ovid). This explains why hermaphrodites on birth were thrown into the sea in Greece and into the Tiber in Rome.[130]

In addition, the book of Genesis recalls that the division of the sexes occurs as a result of divine will. The existence of the hermaphrodite constitutes an act of rebellion against this state of affairs and thus establishes its proximity to the evil practice of sodomy. In primitive Christianity (for example in Clement of Alexandria), sodomy and hermaphroditism were associated.[131] In the same way that in the genesis of monsters the sins of luxury or satanic intervention were often present (for example, the monster as a result of fornication with a beast or an incubus or succubus), the proximity of hermaphroditism with sodomy was often argued.

Such values are widespread in the European literature on prodigies of the 1500s in the context of religious reform. In this period, Catholics and Protestants both signalled the imminent arrival of the Antichrist and interpreted monsters and hermaphrodites as presages of an apocalyptic occurrence. In addition to the widely distributed texts of Rueff (*De Conceptu et Generatione Hominis*, 1554) and Lycosthenes (*Prodigiorum ac Ostentorum Chronicon*, 1557),[132] in Spain we see *El sumario de las maravillas y espantables cosas que en el mundo han acontecido* (1524) by Alvar Gutiérrez de Torres. Drawing on a tradition that went back to antiquity via Isidore of Seville and St Augustine, this author emphasizes the sinful condition and the ominous aura that surrounds the monster.[133]

The first part of *El sumario de maravillas* examines the presence of different prodigies which are linked to particular historical events in the ancient world, in Christianity and in the Spanish realm. The second part focuses on questions pertaining to astrology. Among the many portents discussed – from monstrous births to earthquakes and hurricanes – two particularly related to the transgression of sexual boundaries are mentioned. The first, taken from Augustine's *City of God*, alludes to sex changes that took place during the mandate of Emperor Antoninus Pius. These metamorphoses announced future calamities: 'y las mujeres y gallinas que en hombres y en gallos fueron convertidos' (and women and hens were become men and cockerels).[134] The second episode comes from the time of Rodrigo Díaz de Vivar, shortly before the Oath of St Agatha taken by Alfonso VI; it describes the birth of a baby boy with two heads 'y en el sexo y natura era doblado' (and the sex and nature were doubled).[135]

The same association between prodigies, sexual ambiguity and bad tidings was present in the context of the famous monstrous birth at Ravenna in March 1512.[136] This was an aberrant birth in many senses – one of them was its condition as a hermaphrodite. Eighteen days after the birth a coalition of papal troops from France and Spain took Ravenna and sacked it. The news of this monster spread rapidly throughout Europe. It was commented upon in Spain too and it left a long-lasting mark. Nearly ninety years afterwards Mateo Alemán discussed it as combining 'los dos sexos, sodomía y bestial bruteza: de todos los cuales vicios abundaba por entonces toda Italia' (the two sexes, sodomy and bestial brutality: of all the vices that abounded at that time in Italy), to be punished by God.[137]

Mateo Alemán associates the hermaphrodite with a monster which brought bad tidings and as a punishment for sodomy. Similar characteristics are present in an event from 1688. News of a birth in Madrid detailed a monstrous baby that possessed both 'natures'. The boy nature was placed in the middle of the face while the female nature was in its habitual place and there were no eyes and the nose was absent.[138] Although the author did not dare to delve into the mysterious realm of 'cosas reservadas al Altísimo' (things reserved to the Almighty), he suggested that this birth could well have been a punishment for the practising of acts against nature. In this way, the birth would constitute a moral warning for all Christians.[139] It has also been said that the monstrous birth may have had political significance, a warning about the concern growing under the reign of the weak and sickly Carlos II as the Spanish Empire waned. The hermaphrodite announced the death of the sovereign and the fall of the Empire.[140]

The same kinds of concern about sin *contra natura* are expressed in refrains, art and the abundant iconography of the sixteenth and seventeeth centuries (in addition to the medical literature) on the subject of bearded women.[141] These women were understood to be manly women of warm complexion and thus capable of turning nutritional material into hair, just as men did. These would be *viragines*, luxurious women who gave themselves over to sin. They fell into the category of 'anti-virgins', presented as creatures who were 'siniestras y de mal agüero' (sinister and of bad omen).[142]

As has already been suggested, the link between sodomy and sexual ambiguity in medical texts was very frequent (see Fragoso, Huarte de San Juan and Andrés de León). The proclivity of 'hombres mariosos' or 'mariones' towards the nefarious crime developed from a physiological cause. These were individuals who had been born as girls but who had suffered an excess of heat in the womb and therefore had changed into men. They conserved certain traits from their female past such as the inclination towards counter-natural sin. In this way the thesis of the 'effeminate sodomite', although not widely current at the end of the seventeenth century, had some weight a century earlier, at least in the Hispanic world.[143]

By associating sexual ambiguity with sodomy, the civil and ecclesiastical authorities at the height of the Counter-Reformation and the move to establish holy matrimony in all social spheres would severely punish any transgression at the borderlands of the sexes. Such an association made all intermediary figures possible victims of the death penalty. An examination of the controls set in place with respect to the limits between the sexes in the sixteenth and seventeenth centuries show how such prohibitions against hermaphrodites operated.

Any kind of retrospective utopian designs must be rejected here. Hermaphrodites at this time did not live in some kind of Arcadia whereby they could elect the sex of their choice. This idea, fuelled by certain rather free interpretations of Foucault's work, should be rejected.[144] The legislation operative in the Hispanic world, at least from the *Partidas* of Alfonso X onwards, drew on the long history of Roman law. This legal framework, as the *Digest* shows, saw the hermaphrodite as a figure that was perfectly compatible with the natural order. But since active participation in civil life required the possessing of one sex and one sex alone (the status of the male entailed huge benefits not open to women) sex had to be assigned to hermaphrodites. In classical Roman law the procedure is not clear.[145] However, some legal texts of the twelfth century do point out that the 'predominant' sex is the one that should prevail.[146] The criteria of the Alphonsine *Partidas* follow suit.[147]

Between the sixteenth and seventeenth centuries the renowned expert in canon law Tomás Sánchez (1550–1610) set out the legal proceedings with respect to hermaphrodites. Drawing on the old *Lex Repentudarum* by Ulpian, he admitted the possibility of the birth of hermaphrodites. In these cases, after due examination by doctors and midwives, the newly born would be declared in accordance with the sex that predominated. Only in those cases where the predominant sex was unclear was the decision left to the parents. A final decision could be made by the individual him- or herself at the onset of the marriage age.[148] However, once a decision had been taken the individual could not change it and could not attempt intercourse with persons of the sex that had been chosen. If this did take place sodomy charges would ensue.[149] Although this ceremony of choice gave rise to numerous theoretical disquisitions,[150] it was a completely exceptional procedure reserved for extremely rare cases of perfect hermaphrodites.[151] What usually took place in cases of doubtful sex was that doctors and midwives would identify the 'predominant' sex of the individual. In the Hispanic realms there existed a lengthy tradition, at least from the fourteenth century, that allowed doctors to function as probate and public advisers in judicial decisions.[152]

As we have said before, during the old regime when sex was viewed as a status marker the legal problem with respect to hermaphrodites was not to determine the 'true identity' of the person but to assign rights and privileges in accordance with sex. Could a hermaphrodite take on the role of chancellor of a university, a magistrate, a judge or a lawyer, Alonso Carranza was to ask.[153] The theologian Martín Azpilcueta (1493–1586), for his part, reminded his readers that the hermaphrodite needed to obtain a dispensation from the Pope in order to be ordained a priest.[154] A similar procedure occurred in those individuals who underwent a change of sex and wanted to change state. Antonio de Fuentelapeña discussed the case of a young girl who changed sex and who separated from her husband. Alexander VI allowed her (now him) to marry a woman.[155] In cases such as the nun of Santo Domingo del Real (Madrid, c.1576), who became a man and then became a friar,[156] Catalina de Erauso (c.1629),[157] who assumed the role of a soldier and Fernanda Fernández,[158] a nun from the Convento de Capuchinas de Granada, virilized in 1792 and freed from her vows, the juridical or civil authorities always appear as mediators which permit any dispensation. This was the way that any possible flow of sex changes might be held in check. In addition, this rather frightening world was also associated with the practice of sodomy in the Counter-Reformation.

Despite what we have said, it would not be correct to argue that any 'sexual nomadism' in *ancien régime* Spain navigated only between events deemed marvellous or 'praeternatural', on the one hand, and condemnation and horror as counter-natural sins on the other. There was a kind of marginal space between these extremes occupied by the hermaphrodite or sex change as *miraculus*, as a sign of redemption. This position could be understood in a literal sense – the providential intervention of God in order to produce a sexual metamorphosis – or in an allegorical sense whereby the androgyne was understood as a symbol of original perfection or lack of difference, an emblem of the harmonious fusion of the opposites. In both cases, medieval and Renaissance thought could rely on a certain reading of the classics, whether of Ovid or Plato. An example of cases in the literal sense would be instances of divine intervention in order to save female saints from rape by having them sprout beards or by converting suddenly into men.[159] The second scenario, in an allegorical sense, found its home in a mystical and esoteric tradition that starts with the revelation of Hermes Trismegistus and the Gnostics (the androgyny of Adam before the Fall and androgyny as a symbol of Christ). Such connections only survived in the Spain of the Counter-Reformation in the ambit of the heresy of alchemy.

Sexual transformation could also be the result of an act of Providence that was loaded with portentous and salvationist significance. This is the Christian tradition associated with a line of saints such as the Portuguese St Wilgefortis or Uncumber and the Spanish St Paula de Ávila.[160] These saints grew beards when about to be possessed by men who were to violate their exclusive devotion to Christ. The transformation in these cases came about as result of the petition of the woman who pleaded with God to change and shake off her tormentors. In the case of St Liberada it was a matter of an imposed marriage by her father.[161]

The cult of these virilized women was alive and well in the fifteenth and sixteenth centuries, decaying during the Counter-Reformation.[162] The Castilian Paula,[163] a villager from Cardeñosa, two leagues from Ávila, would trek almost daily to the hermitage of San Segundo, patron saint of the town, in order to pray. She was accosted by a man who wished to force himself on her. To this unwanted advance the woman responded 'in a virile manner' and repulsed him. On another occasion while walking outside of the town she was once more followed by the man who had been hunting and sought refuge in the rural hermitage of San Lorenzo. St Paula then asked God 'la diese alguna fealdad en el rostro' (to give her some ugliness on her face) and thick hairs sprouted from her chin thus chasing off her assailant.

This episode, related in different versions up to the end of the seventeenth century, was beset by contrasts: the difference between being born in a village and a town and a humble peasant birth and the elevated status of the *caballero*. It has been pointed out that most versions focus not on the virginity of the woman but on her tenacious virility.[164] It has also been suggested that the determination of the Counter-Reformation to reinforce boundaries (heretic/Catholic, old and new Christian, marriage/cohabitation, nature/counter-nature, man/woman) would explain the decline of the cult of St Paula after the implementation of the Tridentine decrees. However, the 'improvement' in sex that was miraculously brought about in the damsel of Cardeñosa implied a too great relaxation of the borders between the sexes. The success of the cult of St Águeda, a clearly feminine figure whose breasts had been cut off through martyrdom, would replace to a large degree the bearded Paula. The first testimonies that indicate such a change come from 1595, when in the San Lorenzo hermitage, partly dedicated to St Paula, an image of St Águeda was accorded a predominant position.[165]

It would be tempting to relate the resistance of the Counter-Reformation to sexual transmutation to the questioning of the same kind of metamorphosis in the natural world shown in medical and demonological texts of the seventeenth century. These texts, as we have seen, were not well disposed towards sexual transformation and defended sexual difference between men and women. But there were also severe legal cautions on the subject of hermaphroditism in the sixteenth century, which often associated sexual ambiguity with sodomy. Any contrast between a supposedly tolerant pre-Tridentine sixteenth century and the mentality of a punitive seventeenth century with respect to sexual ambiguity is hard to uphold.

A different case of miraculous metamorphosis, in this case from the popular repertory of the 'cantares de ciego' from the sixteenth and seventeenth centuries, can be seen in the *Casamiento entre dos Damas*. Here, the beautiful Princess Doña Gertrudis, of the Imperial Court of Vienna, was to receive a *billet doux* from a pretender. Because of an outstanding matter with two no-gooders the lover failed to turn up for the date and fled the city. Doña Gertrudis decides to go after him and crosses several countries ending up in the Hellenic lands. Here she passed as a student and dressed as a man. Under the name Carlos, she managed to find a post as secretary to the prince of the locality. The prince's daughter was an outstanding beauty. The princess fell quickly in love with Carlos. The father, duly pleased, gave his consent for the marriage and merriment was had by all. On his nuptial night Carlos told the princess he was in fact a woman. Both agreed to keep this secret and in this manner they spent four happy years of marriage. Given the number of rumours that began to circulate as to the true sex of Carlos, she underwent several examinations but passed all. Finally, God heard her prayers and, in her encounter with a unicorn, she was violently knocked over and male genitalia emerged as a result. The couple married once more secretly and '[p]asados algunos meses el Cielo les ha dotado en darles un sucesor' (once some months had passed Heaven had rewarded them a successor).[166]

The contact with the unicorn sent by God is what now brings about the change of sex. This animal, of potent medieval significance and often associated with luxuriousness, in the modern age underwent an important transformation. The unicorn became a symbol of Christ. Despite this, in the case of Carlos and Palas, the unicorn maintained the connotation of fecundity. God's intervention was not in order to defend the virginity of the woman Gertrudis but to allow her conversion into a suitable husband in order to continue the lineage of the prince. Rather than a myth of purity as in the hagiographies of saints, what we have here is the myth of fertility. The intervention of Providence not only counters a virtually sacrilegious myth but restores the continuation of the blood line.

In reality, the message of the 'romance de ciego' is the same as that contained in the comedies and the novels of the Siglo de Oro. It was a matter of playing with an attempted fraud in status – a king dressed as a beggar, a noble in the dress of a villein, a woman disguised as a man[167] – that upset the established order. The latter, however, at the end of the store would be restored in triumphal and gracious manner.[168] The confusion arising in *Casamiento entre dos damas* concerns not only sex (Gertrudis appearing to be Carlos) but also social status: a princess disguised as a student and then as a servant. In the end, thanks to a miracle, the problem that drives the whole story, that of the search for ideal marriage, is happily resolved. Those of high status are married with the same and man is united to woman in order to raise offspring that will guarantee the future of the line and the consolidation of alliances.

The miraculous conversion of women into men could also transcend individual accounts in order to be raised to the status of a theological allegory. On the final resurrection all fortunate bodies will gain the status of glory, that is, with no hint of imperfection. This understanding, growing out of the one-sex model, meant that women, as incomplete men, would finally be transformed into men.

A long theological tradition underlined this happening. In the apocryphal book of Thomas, dated from the first or second century, the Virgin Mary was transformed into a man by Jesus in order to find salvation. In the story of the martyrdom suffered by Perpetua and Felicitas, a third-century text recorded in the *Acts of the Christian Martyrs*, the first becomes a male and thus accedes to Paradise.[169] The same possibility is present in the thought of Francesc de Eiximenis in his *Lo llibre de les dones*, translated into Castilian Spanish in 1542 (the original Catalan edition is from 1495). Despite the differences between men and women on the earth, in glory the differences will fade away as perfection (maleness) is found: 'y como en la especie humana la mayor perfección sea la del varón, síguese que las mugeres perderán su forma, serán restituidas a la mayor dignidad y nobleza de la especie humana, que es la de varón' (and as in the human species the greatest perfection is that of man, it is to occur that women will lose their form and will be reattributed the greatest dignity and nobleness that human kind has, that of the male).[170]

The collective metamorphosis evoked by the theologian allows us to make one more reference to sexual ambiguity in the sphere of the miraculous: the use of the androgynous as a symbol of reconciliation of opposing values by alchemy. As is well known, the division of the masculine from the feminine traverses the whole of alchemical thought. Alchemy is not merely an intellectual stance on the world; it involves the art of initiation, a complex 'technology of the self' in which uses of materials and elements are at the same time steps on the way towards spiritual ascension whereby the follower unites his or her soul with God in what alchemists called the 'Great Work'.[171]

In this universe of meanings dominated by the difference between man and woman (and between sulphur and mercury, sun and moon and celestial woman and the father) the image of the androgyne represents the unity of the opposites, the *coincidentia oppositorum*.[172] It represents the primordial sameness of material in its equivalence to gold, which results from the fusion between sulphur and mercury, the symbol of universal transmutation of metals. In the life of the initiate, it represents the unity of the soul with God. This salvationist value enshrined in androgyny affects not only individual liberation but also the emancipation of the whole of humanity. Humanity is represented as prostrate, chained to immediacy and therefore to that which is fragmentary and passing. The return to God, to primordial unity, is achieved through the rejection of immediate experience in order to gain superior wisdom. This is what alchemy offers, a kind of *gnosis* in the strict sense of the word, which demands the death of the opposites and their fusion or marriage in order to give way to the noble, innocent soul represented by the androgyne.[173]

In this sense, the Primordial Man who represents the overcoming of sexual duality in the tradition of alchemy has much in common with Gnosticism.[174] Gnosticism values the androgynous condition of Adam, as in the apocryphal book of Thomas, and that of Christ.[175] The distinction between the sexes would therefore be a consequence of sin; Christ, anticipating an *eschaton* that would abolish all difference, would perish as a male and be resurrected as an androgyne. As has been observed, this tradition does not end with the Gnostics – it continues in the Middle Ages in the work of authors such as John Scotus Erigena,[176] and is revived amongst the Neoplatonists (Marsilio Ficino),[177] and theosophists (Jakob Böhme) of the Renaissance.[178] Alchemy takes part in this revival as Paracelsus' work shows,[179] and Spain is not untouched by this development. Spain, after the brilliant era of Llull, undergoes a period of intense renovation in the mid-sixteenth century.[180] De Centellas refers explicitly to the 'marriage' between the 'dama que mora en el cielo' (lady that lives in the sky) and the 'otra cosa' (other thing),[181] whose fusion would give rise to the 'hijo más noble y singular' (most noble and unusual offspring), in the mode of the original androgyne.

Notes

1 On the science of history as a 'science of the unconscious', see Bourdieu (1982: 10).
2 What historians have achieved in this field in the last twenty-five years can be judged by recalling the words of Jacques Revel (1982: 53–4): 'it is not at all surprising if in societies that are libertarian and conflictual at the same time such as our own, the emblematic figure, the symbol of the hermaphrodite, accrues a new significance. The social and cultural function of these social representations of the intermediate is still to be studied, however.' For *ancien régime* Spain see Vázquez García and Moreno Mengíbar (1995a; 1997a: 185–204; 2000); De la Flor (1999); Cátedra (2001); De la Pascua (2003); Salamanca Ballesteros (2007).
3 The first publication by Foucault on the question of Herculine Barbin is Foucault (1978) on the basis of a reading of a psychiatric report

from the mid-nineteenth century. Cases from the sixteenth to the eighteenth century are discussed in Foucault (1994a: 51–74). See also Laqueur (1990).

4   A good summary of the debate in the mid-1990s is Nederman and True (1996: 498, n. 5). See also the debate between Stolberg, Schiebinger and Laqueur in the journal *Isis*, referred to in Chapter 1.
5   We use this concept in the same way as Passeron (2006: 125–68) who considers history and sociology as epistemologically equivalent disciplines.
6   An example of this kind of error in a study of the 'hermaphrodite' Helena de Céspedes in the sixteenth centiury would be Escamilla (1985).
7   This distinction between discursive levels has been used in Vázquez García (2008).
8   For an example relevant to our subject see Salamanca Ballesteros (2007: 283–312).
9   Foucault (1994b: 116).
10 See Laqueur (1990: 124). Nederman and True (1996: 516) discuss the utopianism of Foucault on this point. Davidson (1987: 19) critiques a different question: Foucault's account of Herculine Barbin implies a 'brevity [which] simplifies the complex relations between the legal, religious, and medical treatment of hermaphroditism in the Middle Ages and Renaissance'. See also Daston and Park (1985).
11 This representation of the hermaphrodite as a 'third sex' is what Nederman and True (1996) consider to be prevalent in the Christian West in the twelfth century. The hypothesis of the third sex as a transcultural phenomenon has been defended by some anthropologists such as Herdt (1994). Fausto-Sterling (2000: 78–114) argues for at least five sexes in the human species.
12 See, for example, the ways in which Laqueur (1990: 28–33) minimizes the differences of Aristotelianism with respect to the 'one-sex' model and how he tries to present as support for his interpretation the medical belief in 'menstruating men' (Laqueur, 2003: 305). While Laqueur argues that in the one-sex model the male body is the paradigm from which the female follows, for Gianna Pomata the existence of menstruating males would imply the contrary interpretation. See Pomata (1992: 56). For his part, Michael Stolberg (2003) appears to maintain the thesis that the single-sex model was replaced by the dual-sex model but argues that this transformation in fact took place later, between the sixteenth and the seventeenth centuries.
13 Cadden (1993: 3). A further example of resistance to 'heuristic simplifications' can be found in Park (1997: 174).
14 Cadden (1993).
15 Daston and Park (1995, 1996, 1998: 203); Park (1997: 179–87).
16 In a somewhat confusing manner, Mary Elizabeth Perry (1999: 411) seems to suggest that at this time biology was considered to constitute a fixed destiny as there was no concept of 'gender': 'most people of this time lacked any concept of a socially constructed gender, and

most believed in sex as an essential quality granted at birth, integral to the "natural order", and essential to a sexually dichotomized, hierarchical and patriarchal sociopolitical system'.

17 On these three cases, see respectively: Perry (1999); Burshatin (1999); De la Pascua (2003).
18 This arrangement is discussed by Laqueur (1990: 135 and 137–8). On clothing as distinctive of rank in this society see Lalinde Abadía (1986).
19 The otherwise excellent piece by De la Pascua Sánchez on the case of María Fernández at the end of the eighteenth century falls into this trap: 'in the story I cover here, the biological elements overlap and desire becomes the protagonist and guide in the construction of a new sexual identity. This fact implies modernity ... because the body appears not as something "given" naturally but as something explored as a result of sexual desire' (439).
20 This does not mean that we reject entirely the category 'gender' when analysing transgressions against 'sex as status' in this period. But 'gender', in inverted commas, would have to be understood in the sense employed by Judith Butler, that is, as something that includes 'sex' and 'sexuality' as constituted by *body performances*. 'Gender', then, forms part of regimes of historically changing relations. These regimes would include the performative acts and their transgressions in any specific historical period. On this 'resignification' or subversive reiteration of 'gender' see Butler (1993: 122–4).
21 Vázquez García (1999: 222–3).
22 Cf. Espósito (2006: 88–9). The idea of 'bare life' is taken from Agamben (1998).
23 Foucault (1990a: 121).
24 Elias (1983: 255).
25 Elias (1983: 256), and Davis (1984: 58).
26 See Laqueur (1990: 136–7) on the deliberations of the judges on the case of Marie de Marcis at the beginning of the seventeenth century.
27 We have termed this complex the 'regime of true rank' in Vázquez García (2008).
28 Bakhtin (1987: 273–331).
29 Laqueur (1990: 125) correctly points out that in the absence of any system that fixed the sexes on a biological basis, institutions attempted to consolidate difference on the basis of severe punishment of any transgression. However, his work centres on medical accounts, on travel literature and stories of marvels and hardly touches upon sources of sexual ambiguity as a sign of sin or negative portent. On transvestism as a challenge to heteronormativity see Butler (1993: 124–37).
30 We take this three-dimensional structure (*mirabilis, magicus, miraculus*) from the work of Le Goff (1992: 27–44). On the concept of *mirabilis*, cf. the excellent piece by Park (2000).
31 Nederman and True (1996: 501); Daston and Park (1998: 49); Vázquez García and Moreno Mengíbar (1997a: 187–8).

32 Kappler (1986: 334-5).
33 De Granada (1989: 162) writes of the 'hermosura de las cosas que por la divina Providencia ... fabricadas' (beauty of the things that ... have been fabricated by divine Providence), including the roundness of the earth, the flowers and trees, which in 'su grande variedad nos son causa de un insaciable gusto y deleite' [their great variety are cause in us of insatiable pleasure and delight].
34 Canguilhem (1980: 178–9) locates the beginning of this process in the 'mechanistic physics and philosophy', as a contrast to Renaissance bestiaries. Park and Daston (1981: 36–7) also follow this teleological argument. Later on, in their *Wonders and the Order of Nature* (1998: 176), they rectify this error.
35 Daston and Park (1998: 175).
36 The text by Gutiérrez de Torres (1952) falls into this category. Despite what the first part of its title might imply, this book refers to ominous prodigies as divine signs of calamities to come. The first part of the book interprets numerous historical events, including some in the history of Spain, in accordance with this perspective. The second part is a kind of treatise on astrology. For this reason, we include it in the field of *magicus* rather than that of *mirabilis*.
37 The literature on marvels (*maravillas*) does not exhaust the field of the *mirabilis*. Belonging to this set are also the 'cámaras de maravillas' (from the German *Wunderkammer*). These exhibited, amongst other things, monsters and portents. In Spain they were first found in collections during the reign of Felipe II and included, in addition to arms and tapestries, paintings representing these beings. Amongst these was the portrait of Brígida de Peñaranda, the famous bearded lady, painted by Sánchez Cotán in 1590. Brígida would exhibit herself in exchange for money. See Salamanca Ballesteros (2007: 253–79). On the interest in monstrosity at the courts of Spain and of the Austrias, see Bouza (1991). On the 'curiosities of Nature' included in the 'cámara' of Juan de Lastanosa (1607–84), the *hidalgo* friend of Baltasar Gracián, see Correa Calderón (1961: 26–32).
38 The eighteenth-century De Arreaga (1746) belongs to this genre.
39 Mexía (1989: 502); De Torquemada (1982: 105–6); De Fuentelapeña (1978: 80).
40 De Torquemada (1982: 116–17). The same reference to the testimony of Calliphanes on the Nasamones appears in Covarrubias (1979: 531). Prodigies and portents described in the *Silva de varia lección* (1540) by Pedro de Mexía (1989: 238) do not include episodes of sex change. However, two events portraying women who change dress and take on male roles of pope (the famous Pope Joan) and emperor are mentioned. 'Amazon' women are also discussed (244–61). Finally, the case of Heliogabalus who tried to operate on himself to become a woman is mentioned (713).
41 'unas mugeres se forman un poco hazia la parte derecha de la madre, que llamamos varoniles; y aquestas son más calientes que no las otras

mugeres, pero son menos calientes que los hombres y por esto crían barva. Pero tienen menos barva que los hombres' (some women are formed a little towards the right side of the mother; we call these mannish. These are warmer than other women but not as warm as men and this is why they grow beards. But they have less beard than men) (De Fuentes, 1547). We quote from Sanz Hermida (1997: 1005–6).

42 On the 'natural' condition of hermaphrodites and sex changes, see De Torquemada (1982: 116 and 187–90); De Fuentelapeña (1978: 181, 242); on changes in sex as natural events see Pérez de Moya (1585: 263); De Pineda (1589: 109^r); De la Cerda (1599: 518); Martín del Río (1991: 395); Nieremberg (1643: 54–5) and De Torreblanca y Villalpando (1678: liber II, cap. XVII). De Fuentelapeña also admits the existence of menstruating men although he explains them by stating that they are in reality 'hidden hermaphrodites' who have both natures, one internal and the other external (De Fuentelapeña, 1978: 230). The texts by Martín del Río and Torreblanca y Villalpando can be considered treatises on demonology. However, their distinctiveness from the literature of marvels is unclear in many respects.
43 Martín del Río (1991: 393) and De la Cerda (1599: 519).
44 De Torquemada (1982: 190).
45 Nieremberg (1643: 54–5). The imagination of parents was often invoked as a cause of monstrous births (De Fuentelapeña, 1978: 170–1).
46 De Pineda (1589: 163^v) discusses how the warmer female is still, according to Aristotle, less of a male than a full male.
47 Martín del Río (1991: 395) wrote that Galen and Avicenna and others believed that 'la mujer es una especie de varón monstruoso e imperfecto' (woman is a kind of imperfect and monstrous male); Fuentelapeña (1978: 244) that 'es más frecuente la mutación de muger en hombre que la de hombre en mugger ... porque la naturaleza siempre aspira a lo más perfecto' (the mutation of woman towards man is more common than man towards women ... because nature always aspires towards the most perfect). Pineda discounted stories of men becoming women (De Pineda, 1589: 109^r)
48 Stolberg (2003).
49 Fuentelapeña (1978: 181) agrees with Aristotle that the hermaphrodite is a monster but not in the same way. Instead of referring to a deviation or sin of nature, De Fuentelapeña prefers to refer to the hermaphrodite as 'extraviarse de lo ordinario' (moving away from the ordinary). For this reason, the hermaphrodite ('sexo hermafrodítico') with two sexes is not as monstrous 'como el sexo que llaman neutro', that is, the undefined sex.
50 De Fuentelapeña (1978: 181) describes generation as a battle between male and female semen; the hermaphrodite results from an undecisive battle in which neither 'pueda vencer y consumir a la otra' (is able to defeat and consume the other). He even believes that the hermaphrodite can fertilize him- or herself (pp. 229–41).

51 Ibid., pp. 247–8.
52 Ibid., p. 248.
53 Martín del Río (1991: 395).
54 Ibid.
55 Ibid.
56 Ibid., p. 196.
57 Ibid.
58 The *Relación* dates from 1617, although the event must have taken place before 1613 (cf. Morel d'Arleux, 1996: 268).
59 Cf. Ettinghausen (1995: 12).
60 On the *relaciones* of prodigies as an instrument of social control wielded by the Church and the authorities see Ettinghausen (1995: 14; 1993).
61 Facsimile edn of the *relación* in Ettinghausen (1995) with a résumé on pp. 36–7. The case is commented upon extensively by Morel d'Arleux (1996: 263–8).
62 The accompanying symbolism is examined minutely by Morel d'Arleux (1996: 265–6). Ettinghausen (1995: 37) also suggests the allusion to hermaphroditism but interprets the case in accordance with modern scientific language. He writes: 'it would seem to be a case of hermaphroditism or rather of mixed gonadal disgenesis'.
63 'The anatomical work of Galen was sufficiently adequate to fulfil the intellectual demands of of anatomists over fourteen centuries, during which the history of anatomy is one continuous line of exposition of the Galenic point of view' (Alberti López, 1948: 42–3). Others write: 'Galen's work reaches its maximum heights in the sixteenth century when numerous Latin versions were printed throughout Europe ... Despite some highly valid studies, we still do not possess a complete panorama of Galenism and Spanish medical humanism for the sixteenth century' (Riera Palmero, 2001: 166).
64 On the polemic between Spanish Renaissance medicine and the Arabic medical tradition, see García-Ballester (1984: 19–46). See also García-Ballester (2002).
65 Granjel (1980: 32) observes that 'Hippocratic' Galenism evident in the translations commented on by Mena, Cristóbal de Vega and Valles bears witness to the orientation of medical humanism in the second half of the sixteenth century. This made it possible to convert Hippocratic texts into the model of medical knowledge and practice without questioning the authority of Galen (Granjel, 1980: 32). López Piñero et al. (1983c: 392) note 'Valles was, in reality, one of the main European figures in a tendency that emerged in the humanist current. This tendency, without questioning Galen and the authority of his system, transformed Hippocrates into the principal reference for medical knowledge and practice. Such a position, which we can call 'Hippocratic' Galenism drew on Hippocratic texts through the lens of humanism'. Pérez Ibáñez (1997: 114–15) writes 'Renaissance medicine returns to the texts of Hippocrates and Galen as primary sources ... Hippocrates and Galen are the two most cited authors in

[medical] compositions. The quotations are precise and may incorporate references to the works of textual critique to which [the original texts] were submitted'.
66 Montaña de Monserrate (1551: fol. 61ᵛ) writes 'la mujer es diferente del varón' (woman is different from man) and this is why the genitalia of men are outside the body, due to greater heat.
67 Valverde de Amusco (1556: 65ᵛ) acknowledged that woman's parts were inside the body in order for procreation to take place. Valverde was a disciple of Realdo Colombo, whose master was Vesalio. Thomas Laqueur reproduces some of the laminates that appeared in the work by Valverde – which were renowned in his time – in order to show how dominant the one-sex model was (cf. Laqueur, 1990: 76–7).
68 Fragoso (1570: 26ᵛ) also acknowledged that nature had made men's parts outside of their bodies as they were 'más calientes en la composición' (warmer in their composition).
69 De León (1590: 28ᵛ) focused on the similarities: 'las mujeres también tienen testículos y miembro viril, aunque ocultos y escondidos, y redondos como los de los hombres' (women also possess testicles and virile member, although they are hidden and covered and round like those of men).
70 Álvarez de Miravall (1597: 286ʳ): 'Dicen los filósofos que la mujer no es otra cosa sino un varón imperfecto' (The philosophers say that woman is nothing but an imperfect man). On the importance of this text, see Granjel (1967: 93–116).
71 Huarte de San Juan (1976: 315).
72 Ibid., p. 328 confirmed that seed ('simientes') from both male and female was required for procreation.
73 Lobera de Ávila (1551: 58ʳ). Lobera de Ávila asserts that women who usually sleep on the right-hand side rarely give birth to girl babies (38ᵛ). Luis Lobera de Ávila (*fl.* 1530) was a doctor in the service of Carlos V (Alberti López, 1948: 61–7). Huarte de San Juan, for his part (1976: 333–6) offers advice on how to conceive male children.
74 Lobera de Ávila (1551: 33ᵛ). Álvarez de Miravall (1597: 280ʳ–280ᵛ) writes that women have an 'expultrix' facility with respect to semen, situated in the testicles, and an 'attractrix' impulse (absent in males) with respect to male semen once in the womb.
75 Montaña de Monserrate (1551: 65ʳ–66ᵛ) wrote 'la simiente de la mujer no tiene nunca aquella perfección de adherencia y viscosidad ... lo cual consta por experiencia pues que para la generación del cuerpo humano es tan necesaria la simiente del varón que no se puede engendrar sin ella' (the seed of woman never possesses that perfection of adherence and viscosity ... which is shown by experience since the male seed is so necessary that procreation is not possible without it). Montaña opposed the Hippocratic division of the womb into cavities and believed that the womb only had a 'very small cavity' (72ʳ). According to historians Jacquart and Thomasset (1989: 147–9) this belief in the cavities of the uterus was derived from the idea of the female womb that certain animals possessed, in particular pigs.

76 Montaña de Monserrate defended this notion: the foetus was formed by the fermentation of the semen (mainly male but not excluding the female) with menstrual blood. With arterial blood the solid parts of the body were formed and with the blood of the veins the softer parts were formed (Montaña, 1551: 61^v–62^r). This distinction goes back to Alî Ibn al-Abbas (Iglesias Aparicio, 2007: 64).

77 Valverde de Amusco (1556: 65^r) also wrote that the male seed was the most important in the process.

78 Both Huarte de San Juan (1976: 336) and Peramato (1576: 67) discussed the female-like nature of many men amongst the Scythians.

79 The Portuguese Pedro de Peramato (fl. 1575), educated at the University of Salamanca, classified hermaphrodites according to four types. Three were predominantly of the masculine type and one feminine (Peramato, 1576: 117^{r-v}). On Peramato, see Chinchilla (1967: 77–81).

80 'Mas que aya hombres de entrambas naturas y que los haya habido no es cosa de dexar de sabello ... Yo vi en Salamanca un hijo de un caballero, que tenía entrambos miembros de hombre y mujer' (But that there are and have been men of both natures is not out of the question ... I saw in Salamanca the son of a gentleman who had both members of man and woman) (Sánchez Valdés de la Plata, 1598: 130^r.). This text was written between 1545 and 1550. The book of the La Mancha doctor Juan Sánchez Valdés de la Plata, as found by Elena Ronzón, is in part a plagiarized version of the often published *Silva de varia lección*, by Pedro Mexía, which appeared in 1540 (cf. Ronzón, 1998). However, the chapter by Sánchez Valdés de la Plata on sex changes does not appear in the work by Mexía.

81 Luis Mercado, considered to be the 'Saint Thomas of Medicine, was responsible for the systematization in sixteenth-century Spain of traditional medieval medical doctrines' (López Piñero et al., 1983c, vol. 2, 56–9). In ch 6 ('De monstroso conceptu') of the third book of *De Mulierum Affectionibus* hermaphroditism is not examined extensively. It is discussed as an example of monstrosity, together with other forms of the same, in order to show the diversity of this phenomenon (Mercado, 1579: 385).

82 Mercado (1579: 387–94).

83 Ibid., p. 385.

84 Ibid., p. 235.

85 Andrés Laguna (1510–59), doctor to the court of Fernando el Católico and in the service of Cardenal Francisco de Bobadilla y Mendoza from 1545, cared for Popes Paul III and Julius III. In a letter to the latter from 1551 Laguna wrote that he had seen with his own eyes men in Germany that lactated and menstruated. He did not consider them to be monstrous cases (cf. Pomata, 1992: 57). On the career of Laguna, see Granjel (2001: 11–16).

86 Álvarez de Miravall (1597: 278^v). Fragoso (1627: 21) for his part, alluding to what was being rediscovered at the time as the clitoris, although not identifying it as such, wrote of certain women being in the possession of a kind of protuberance (citing Paul of Aegina,

Avicenna and Albucasis as authorities) and who united 'as if they were men'.

87 These were individuals who, in the womb, would have been female for several months. They were transformed into male foetuses through an excess in heat and would retain, on birth, certain forms of behaviour and characteristics of their first sex with a tendency to commit the 'nefarious sin' (Fragoso, 1627: 163). Huarte de San Juan in a work published six years before that of Fragoso had said practically the same thing (see Huarte de San Juan, 1976: 315–16). Fifteen years later, Andrés de León (1590: 28^v) repeated the same idea.

88 As we have stated, Andrés Laguna mentioned that he had seen numerous cases of lactating men. De Fuentelapeña (1978: 253) wrote that Bernardino Montaña de Montserrate had also seen cases of men who breast-fed their children. See also the study by Dr Ángel Pulido (1880) who referred to examples from the sixteenth and seventeenth centuries but with no mention of Spanish doctors.

89 Fragoso (1627: 162–3) devotes a chapter to the question. He admits the possibility of change in both directions. Huarte de San Juan (1976: 315–16) argues the same. Sánchez Valdés de la Plata (1598: 17^v) invokes Aristotelian teleology in order to reject episodes of supposed feminization. Peramato (1576: 117^r) mentions two cases in Spain, seeing them not as sex changes in the true sense as described by Pliny, but as the expulsion of a sex that was hidden. Álvarez de Miravall (1597: 286^r) also casts doubt on cases of feminization: 'Y así leemos que las hembras se muden en varones, y que muy raras veces se ha visto lo contrario, porque siempre la Naturaleza procede de lo más malo a lo mejor, y no al contrario' (And so we read about women who change into men and very rarely have we seen the contrary, because Nature always proceeds from the worst towards the best, and not the reverse). He cites all the Graeco-Roman authorities on the question of masculinization (Hippocrates, Lucinio Murciano, Pliny, Aulo Gelio, Ovid, Livy). De León (1590: 28^v) describes sex changes in both direction but only in the womb.

90 Cf. Park (1997: 179–83).
91 Stolberg (2003).
92 Laqueur (2003: 301–3).
93 See the entry on 'García Carrero, Pedro' by López Piñero et al. (1983a: 374). See also Granjel (1978: 24) on the importance of Pedro García Carrero.
94 García Carrero (1605: 1034).
95 Ibid., p. 1035.
96 Ibid., p. 1181.
97 Daston and Park (1995: 424).
98 In this period, as the *Tesoro de la lengua española* by Sebastián de Covarrubias shows, the two theories were current: the Aristotelian theory that identified the hermaphrodite as a type of monster and the Hippocratic that considered the hermaphrodite as an unusual

but not monstrous product of generation. See Covarrubias (1979: 118).
99 García Carrero (1605: 1181).
100 For example, ibid., pp. 1054, 1115.
101 We have used the 1630 edn with the title *De Partu Naturali et Legitimo*. The importance of Carranza for medicine is shown by the fact that Anastasio Morejón provides an entry on him in Chinchilla (1967: 330–1).
102 Carranza (1630: 645).
103 On the translation of 'peccatum' as an error of nature and not as a human sin in this context see Daston and Park (1998: 201).
104 Carranza (1630: 647) arguing against the thought of Clement of Alexandria.
105 Ibid., pp. 645–6.
106 Bravo de Sobremonte was 'The most representative and prestigious figure of moderate Galenism in Spain in the mid-seventeenth century. Numerous elements of modern doctrines were accepted by him' say López Piñero et al. (1983b: 133). Cf. also Granjel (1967: 134).
107 'Promptuarium XXIV', 'De sexus mutatione' and 'De hermaphroditis' in Bravo de Sobremonte (1671: 246–9, 249–50) and 'Resolutio I. Utrum sexus transmutatio permitatur naturae' in Bravo de Sobremonte (1679: 198).
108 Bravo de Sobremonte (1679: 198). In the same promptuarium, published eight years earlier, those authorities that gave credence to cases of masculinization are mentioned. The reverse case scenario appears to be rejected from the start. Later, citing Luis Mercado and André du Laurens, amongst others, three arguments that refute such a possibility are outlined. Women transformed into men are in reality hermaphrodites whose male members emerge late. They may also be bearded women as a result of menstrual alterations or women with a kind of deformation of their parts ('Promptuarium XXIV', 'De sexus mutatione' in Bravo de Sobremonte, 1679: 248). In this text, only the Madrid and Úbeda cases are mentioned. Both are cited by Peramato and another case of metamorphosis in Villaviciosa (Madrid) is discussed ('Promptuarium XXIV' in Bravo de Sobremonte, 1679: 247).
109 Bravo de Sobremonte (1679: 198).
110 Ibid., p. 197. This thesis, derived from the work of Du Laurens, has been interpreted by Laqueur as a softening of the distinction between men and women and as a result is still firmly within the one-sex paradigm (Laqueur, 2003: 303). Rather, we understand that the point should be related more closely to the context of the discussion (especially in the case of Bravo de Sobremonte); what is being insisted upon is that women are not monsters and that they do not belong to another species.
111 Bravo de Sobremonte (1679: 197).
112 The comparison by ibid., p. 199.

113 Ibid.
114 Ibid., pp. 199–200.
115 Ibid., p. 200; Bravo de Sobremonte (1671: 249).
116 On the attitude of these authors to hermaphroditism, see Park (1997: 178–84).
117 'Promptuarium XXIV', 'De hermaphroditis' in Bravo de Sobremonte (1671: 249).
118 Ibid., p. 250.
119 Around 1632, doctors such as the court physician Juan de Quiñones and Gerónimo de la Huarta had no difficulty in affirming that male Jews menstruated periodically just like women. See Beusterien (1999). One way of according infamy to Jews was precisely by attributing to them the category of 'imperfect males', that is, women. A tract published in Barcelona in 1606 refers to the case of a Granada-born man who, as a result of an act of sorcery, gave birth to a foetus with the form of the devil. On this case, see Córdoba (1987) and Salamanca Ballesteros (2007: 311).
120 Granjel (1978: 149) refers to this work as 'the most important single contribution to the medical literature of the seventeenth century on the subject of congenital anomalies'.
121 Rivilla Bonet (1695).
122 Ibid., p. 8^r.
123 Ibid., p. 9^v.
124 Ibid., p. 14^v.
125 Ibid., pp.35^v–36^r.
126 Sebastián de Covarrubias (1979: 118) also contemplates this cause when he refers to 'andrógenos' or 'ermaphroditos'.
127 Rivilla Bonet (1695: 37^v).
128 The jurist Alonso Carranza severely rejects the Hermetic and Gnostic theories of the first man created by God as a hermaphrodite (Carranza, 1630: 645). On the beliefs of some early Christian sects on Adam and Eve as a 'parted androgyne', see Brown (1988: 268).
129 It was Jacques Derrida who first elaborated the idea of 'resignification' as the reiteration of deconstructive effects. The idea has been most used by queer studies and by Judith Butler. On the need of 'resignifying' without losing the terms 'gender' and 'sex' while realizing that the terms ('race' included) are part of oppressive regimes see Butler (1993: 123): 'precisely because such terms have been produced and constrained within such regimes, they ought to be repeated in directions that reverse and displace their originating terms'.
130 Delcourt (1970: 66–76). On monsters as negative auguries see Salamanca Ballesteros (2007: 171–200).
131 Boswell (1980: 356–7).
132 Cf. Wilson (1993: 65–77); Daston and Park (1998: 183); Salamanca Ballesteros (2007: 171–200).
133 Gutiérrez de Torres (1952: a5).
134 Ibid., p. b6.

135 Ibid., p. f1.
136 Daston and Park (1998: 177–80); Salamanca Ballesteros (2007: 192–4).
137 Alemán (1968: 246). Rivilla Bonet (1695: 37ʳ) also mentions the monster of Ravenna. The deciphering of the significance of monstrous births, as well as the interpretation of certain dreams, could be valuable in terms of political predictions, as in the example of the praise given to Mateo Vázquez de Leca, the secretary of Felipe II, by Calvete de Estrella. In this text a monster born in Ledesma is referred to and is seen as a bad sign for the monarchy: C. V. Calvete de Estrella, 'Elogio del Secretario Real Don Mateo Vázquez de Leca' (orig. c.1585), cited in Díaz Gito (1990: 21).
138 See the *Relación verdadera y caso prodigioso y raro, que ha sucedido en esta Corte el día catorce de mayo de este año de 1688* in Ettinghausen (1995: n.p.): 'la criatura que nació era niño y niña, con dos naturalezas, la de niña en la parte común y la de niño en mitad de la frente' (the child that was born was a boy and a girl, with two natures, that of the girl in its common place and that of the boy in the middle of the forehead).
139 Ibid.: the deformity would serve 'de exemplar a todos los Católicos Christianos, por si en su generación huvo algún excesso vicioso, que suele el Cielo castigar en los hijos travesuras y desacatos de los padres' (as an example to all Catholic Christians, for if there were any vicious excess in the act of generation, Heaven usually visits the misdemeanours and carelessness of the parents on their children).
140 This reading proposed by Morel d'Arleux (1996: 270) is a little forced if we go back to the original text. Here, it is only said that in the court of Carlos II news of these cases from 'Reynos muy distantes' is extensive but none comes from Spain. Obviously, the consumption of this kind of news may well have been quite common, given the precarious nature of the Habsburg monarchy, but to extrapolate a political sign from it appears to be excessive.
141 Fernando R. de la Flor has analysed these sources and indicates that Carlos V had sought from Sánchez Coello, c.1555, two portraits of bearded women, now lost. To these we can add the also lost account by Diego Valentín Díaz of 1660 and, especially, those of José de Ribera ('Magdalena de los Abruzos', 1631) and of Juan Sánchez Cotán ('Brígida del Río', 1590). Brígida del Río was mentioned, as we have said, by Bravo de Sobremonte in 1679. De la Flor (1999: 269–305) points to references from Mateo Alemán in 1599, Covarrubias in 1611 and Jerónimo de Alcalá in 1624. The portrait by Sánchez Cotán was made into numerous copies and other painters also depicted her. Magdalena was exhibited in the court of the Viceroy of Naples.
142 De la Flor (1999: 286–8).
143 Federico Garza has presented this same argument against the interpretations of McIntosh, Bray and Trumbach. See Garza (2002: 96–7). On the 'effeminate sodomites' of Mexico, who were persecuted by

the Inquisition in the mid-seventeenth century, see Garza (2002: 239–48). Evidently, in contrast to the late nineteenth-century 'invert', these sodomites were one variety amongst all kinds of sodomite.

144 This rather utopian understanding has perhaps been encouraged by a superficial reading of texts like that of Ambroise Paré. Paré (1993: 38) refers to having to choose a sex but the need to keep to it under pain of death.

145 However, the *Lex Repentudarum* from the third century CE, drawn up by Ulpian, allowed hermaphrodites to be either men or women according to the sex that predominated. See Darmon (1985: 41).

146 On this point, cf. Nederman and True (1996: 511–15).

147 See Alfonso X El Sabio (1843–4: vol. 4, partida VI, tit. 1, ley 10).

148 'Hermaphroditus' in Sánchez (1629: 339–41) cited in Barbazza (1984: 34–5). See also Darmon (1985: 42).

149 Antonio de Torquemada mentions two cases of hermaphrodites in Burgos and Seville who went back on their chosen sex and were burned for sodomy (De Torquemada, 1976: 116). More surprising and legally problematic is the case of marriage between two hermaphrodites that took place in Valencia in 1662. The case is mentioned by the Valencian jurist Matheu y Sanz in his *Tractatus de Re criminali* (1686), cited in Tomás y Valiente (1990: 54–5).

150 Antonio de Fuentelapeña discusses the process of election in detail in two of his 'doubts': 'Si el hermafrodita, en quien prevalecen con igualdad y perfectamente ambos sexos, podrá a un mismo tiempo casarse con dos, esto es, con un hombre y una muger' (If the hermaphrodite in whom both sexes equally and perfectly prevail could at the same time marry two, that is, a man and a woman) and 'Si ... sucesivamente podrá casar con diversos sexos dicho Hermafrodita' (If ... the said hermaphrodite could marry successively with different sexes), in De Fuentelapeña (1978: 181–6, 187–90). De Fuentelapeña (1978: 187) points out that the pledge to adopt one sex or the other must be made in the presence of the bishop or ecclesiastical judge. The jurist Matheu i Sanz also considers this question in his controversy XLVIII in his *Tractatus de Re criminali* (1686), cited in Tomás y Valiente (1990: 54–5). Alonso Carranza (1630: 646) after classifying hermaphrodites into four groups including in the fourth group those hermaphrodites with the genitalia of both sexes 'completa et perfecta' also refers to this ceremony although, citing Tomás Sánchez, he appears to confine it to the latter class of hermaphrodite.

151 Katharine Park believes that this 'choice', reserved for cases of 'perfect hermaphrodites', was in reality extremely rare. Nevertheless, we have already mentioned the two cases of the hermaphrodites burned at the stake in Burgos and Seville, cited by Antonio de Torquemada. Another case, described in the *Relaciones histórico-geográfico-estadísticas* ordered by Felipe II in 1578, is of a girl from Valdaracete (Madrid province), mentioned below. She was baptized

as Estebanía and led a life as a woman, although she was reputed to have the physical strength of a man. In Granada, where she settled at the age of little over 20, the magistrates of the Chancillería decided to 'ver y examinar por matronas y parteras' (have her seen and examined by matrons and midwives) 'y fue hallada hermafrodita' (and she was found to be a hermaphrodite). Then 'la mandaron que escogiese en el hábito que quería vivir e andar y eligió el de hombre' (they commanded her to choose the habit in which she wished to live and be and she elected to take that of a man); she was married later to another woman (Viñas y Mey and Paz, 1949: 630–1). It would not seem, at least in Spain, that choosing sex was as uncommon as Park has suggested.
152 Cf. McVaugh (1993).
153 Carranza (1630: 646–7).
154 Navarro Martín de Azpilcueta, *De Censuris Ecclesiasticis*, cited in Covarrubias (1979: 119) and Navarro Martín de Azpilcueta, *Num primum in unum quasi corpus coagmentati ... capitum Juris Canonici expositum et Legum Iuris Civilis* (Lyon, 1591) cited in Morel d'Arleux (1996: 263). De Fuentelapeña also refers to the possibility of ordaining hermaphrodites. See De Fuentelapeña (1978: 162–4).
155 De Fuentelapeña (1978: 245).
156 Peramato (1576: fol. 117^{r-v}) and Pérez de Moya (1585: 263). The ecclesiastical authority (it is uncertain as to whether bishop or Pope) authorized the masculinized nun to become a friar.
157 Perry (1999: 408–9). Catalina received permission from Felipe IV to act as a soldier. In 1529 she sought an audience with the Pope who allowed her to dress as a man and carry arms.
158 De la Pascua (2003: 443). In this case it was the bishop, after a medical examination, who freed the subject from her vows.
159 On the possible classical antecedents of the hagiographic discourse on saints who became men see De la Flor (1999: 298).
160 To these examples we should add another hagiographical repertoire, that of 'cross-dressed' saints, who lived as men in order to escape from the family or marital authority and to take up a life of perfection. This group includes Sts Eugenia, Anastasia, Margarita and Pelagia, whose sixth- and seventh-century lives are described in the *Leyenda Dorada*. Cf. Steinberg (2001: 67). The behaviour of St Margarita seems to have inspired Joan of Arc, according to Steinberg (2001: 68).
161 Nieremberg (1643: 55). On this tradition, see Cátedra (2001: 131–44). For the case of St Gala, see De la Flor (1999: 287).
162 Cátedra (2001: 137).
163 We focus on St Paula because she is the major Spanish example of these 'bearded women'. Even so, the image of St Wilgefortis was very popular in Spain in the fifteenth and sixteenth centuries as the extensive iconography showing her marytrdom suggests. One of these representations can be found in the Museo Diocesano of la Seu d'Urgel. Cf. also De la Flor (1999: 298–9). Strangely enough, of St Paula there remain no iconographies from the same centuries.

164　Cátedra (2001: 136).
165　Ibid., p. 133.
166　Compte Masía (2000: 213–14). Thanks go to Chema Fraile for providing a copy of this text.
167　This subject has been studied, amongst others, by Romera Navarro (1934: 269–86); Homero Arjona (1937: 120–45); Bravo Villasante (1955); Castro Pires de Lima (1958); Ashcom (1960: 43–62); McKendrick (1974); Inamoto (1992); Sanz Hermida (1993); Fuchs (1996); Escalonilla López (1998), and Velasco (2001). Dr Lourdes Bueno Pérez, of the University of Austin, possesses an extensive inventory of different types of gender transformation in the Spanish theatre of the Golden Age. Cf. her presentation at the round table on 'Representaciones del Género y de la Identidad Sexual en el Teatro Español') during the summer programme 'El Género del Teatro: Las Artes Escénicas y las Representaciones de la Identidad' (University of Cadiz, 6 July 2007). We thank Dr M. Jesús Ruiz Fernández (University of Cadiz) for this information.
168　On the distinction between forms of transvestism that reaffirm heterosexual culture and those that subvert it, see Butler (1993: 124–37).
169　Cátedra (2001: 136).
170　Eiximenis (1542: 29) cited in Morel (1996: 271). On the relation between illness and sins of luxury in Eiximenis see Solomon (1999).
171　Eslava Galán (1987: 13–47).
172　Authors such as Mircea Eliade (1984) and Jean Libis (2001), by following (erroneously in our view) Jungian theory on archetypes, find the androgyne in these primitive scenarios. A critique of Jungian interpretations, which essentialize symbols without placing them in their specific contexts of domination, is Vázquez García (2001). Frances A. Yates (2006: 91) comments on *The Chemical Wedding*, supposedly written by Christian Rosencreutz, originally published in 1616, as 'an alchemical fantasia, using the fundamental image of elemental fusion, the marriage, the uniting of the *sponsus* and the *sponsa*' (male and female principles).
173　This alchemical symbol of the androgyne transcends strictly alchemical texts. See, for example, the illustrations that accompany the *relación* of the sexual metamorphosis of the nun of Úbeda, published in 1617 (Morel, 1996: 265–6). On the significance of the androgyne in the comedies *El Aquiles* and *La Dama del Olivar* by Tirso de Molina, see Paterson (1993). In *El Aquiles* a man dressed as a woman appears. This is unusual in the context of the Golden Age; the reverse is far more common. Aquiles, imitating the masking of other gods and demi-gods, is disguised as a woman in order to find his beloved Deidamia.
174　Eliade (1984: 129–36); Libis (2001: 58–64); Eslava Galán (1987: 36–7).
175　Meeks (1974).

176 Eliade (1984: 131–2).
177 Libis (2001: 104–5). Pedro Sánchez de Viana (1589: 3), drawing on the Neoplatonists León Hebreo and Marsilio Ficino, alludes to the androgynous condition of Adam only to reject it immediately, cited in Morel (1996: 266).
178 Eliade (1984: 129–30) and Libis (2001: 83–4).
179 On Adam as androgyne in the thought of Paracelsus, see Koyré (1971: 107).
180 On the Valencian Luis de Centellas (or Luis Centelles) (*fl.* 1552), see García Font (1976: 208–12) and Eslava Galán (1987: 114–16).
181 L. de Centellas, *Coplas de Don Luis de Centellas sobre la Piedra Filosofal* (*c.*1552), in De Centellas (1987: 117). This 'lady' would be the feminine principle, associated with mercury, the generating force. Medieval Spanish alchemy would assign mercury an all-important role. However, in the alchemy of the Renaissance this hegemony would be transferred to sulphur, the masculine principle and active power of the solar essence (Eslava Galán, 1987: 31).

Chapter 3
Between Diagnoses: Hermaphroditism, Hypospadias and Pseudo-hermaphroditism, 1870–1905

In 1847 the Mallorca-born Dean of the Paris Faculty of Medicine, Mateo Orfila, wrote that, contrary to the received general opinion, hermaphroditism only existed in the inferior orders of plants and animals and that in reference to human beings 'deberia pues borrarse del lenguage médico la palabra *hermafrodismo*' (the word *hermaphrodism* should be expunged from the medical lexicon).[1] Orfila's remarks constituted a derivation from the ongoing nineteenth-century process of disenchantment with the figure of the hermaphrodite and his or her slow elimination as a real human possibility, to be replaced, as we have seen in the last chapter, by a variety of in-between positions rendered in classificatory systems, following the schemas proposed by influential surgeons and gynaecologists such as Henri Marc, Geoffroy St-Hilaire and Samuel Pozzi.

Orfila's comments indeed illustrate evidence of the changing paradigms of the medicalization of the hermaphrodite in Spain. In this chapter, nevertheless, we emphasize that paradigms did not alter suddenly or systematically across all medical disciplines. Competing accounts explaining the phenomenon of hermaphroditism existed right up to the point where this book leaves off, the 1960s. What we see instead of the steady replacement of one category or classification after another is a contested conjunction of diagnostics, where constant renovation, overlap between cognitive systems and 'ghost' categories survive and continue to inform medical doctors' views on the question. The general consensus that the 'real' hermaphrodite did not in fact exist in the human species gave way to a battle over truth in which new categories (the '*apparent*' or '*pseudo*-hermaphrodite') suggested one state, hermaphroditism, but immediately disavowed it in order to denote *in reality* another, something less than or not quite hermaphroditism. Despite this revision of the free-standing term 'hermaphrodite' and the reliance on a classification that was in itself inherently unstable (how apparent was apparent?), the category continued to emerge surreptitiously as writers harked back to previously rejected classifications. At first sight, it is surprising to note that the notion of the real hermaphrodite would be used in medical reviews not only in the 1920s but also in the early 1960s. A similar process of non-linear accretion of categories and their acceptance, with overlapping terminologies and aetiologies, can be seen in the case of the construction of 'homosexuality' in Spain.[2]

Given these realities, we need to question those accounts which have suggested the constant renovation and step-by-step replacement of medical paradigms on hermaphroditism. Despite the fundamental contribution to this field of study made by Alice D. Dreger, this writer has perhaps overschematized the 'stages' whereby hermaphrodite science was articulated, suggesting a rather too hermetic periodization of diagnostic practices moving from emphasis on visual identification of sex through to reliance on locating and identifying the gonads as the markers of 'true' sex.[3] Other authors would identify a later endocrinological stage in the designation of sex.[4] While Dreger's thought has been re-evaluated in respect of countries such as the United States, Britain and France, a similar critique is pressing for Spain too, although Dreger does not focus on this particular location. It is worth summarizing Dreger's hypothesis briefly.

If previous methods had utilized primarily sight and touch as the principal tools for sex identification – the physician's view and examination of the external genitalia – including a broad variety of characteristics including comportment, hirsuteness, timbre of voice, gestures and the existence of breasts, Dreger argues that from the 1870s onwards the identification of the sex of the individual was determined by reference to the maleness or femaleness of the gonads he or she possessed (that is, the existence of testes or ovaries). This change was consolidated by new early twentieth-century diagnostic techniques that allowed for the extirpation and analysis of the ovaries and testes under the microscope before the patient's death (biopsies).[5] In this way, the gonads and their product (sperm and ova) became the overriding factor in the determination of sex by doctors. In Dreger's words, the 'age of the gonads' had dawned.[6]

This picture soon became more complicated, however. From the early twentieth century, as Nelly Oudshoorn has argued, with the discovery of 'male' and 'female' hormones as secreted by certain organs in the body, hormonal or endocrinological accounts of real sex took centre stage. The 'truth' of the body was now found to lie in the hormones secreted by the body and was not confined necessarily to the presence or otherwise of the gonads but to their function and to the balance effected by their 'male' and 'female' secretions. A further complication arose, however. In the early 1930s it was confirmed that some hormones previously known as 'female' were found in non-hermaphroditic males and vice versa. The idea that the two sexes were radically different and incompatible led, with the aid of new psychological theories of gender development, to the notion that the sexes were a continuity rather than rigidly differentiated. Now, a variety of factors – genital, gonadal, hormonal and psychological – could identify the real sex of the person.

On the one hand, a critique of Dreger's and Oudshoorn's analyses of these broad cognitive changes should not mean that we reject any possibility of tracing historical change in the science of hermaphroditism. On the contrary, for Spain it is indeed possible to identify a number of crucial periods during which new theories of hermaphroditism became influential. On the other hand, the acknowledgement of such periods should not be allowed to obfuscate continuity with older classificatory systems and methods of determining doubtful sex. While the eighteenth century was unarguably the century when the hermaphrodite 'disappears', the 1870s was a period when, in Spain, systems of classification underwent an acute crisis and medical doctors began to rely increasingly on *individual* diagnoses of clinical cases, having argued that it was impossible, given the complexities of the cases and their accurate description, to draw up schemas that satisfied all eventualities.[7] Furthermore, while acknowledging the importance of endocrinological accounts in Spain from the mid-1910s onwards, particularly in the light of the influential theories of Gregorio Marañón, we emphasize the concurrence of other theories of hermaphroditism that existed problematically in some instances alongside these newer frameworks.

Finally, we can detect from 1940 onwards a movement towards the acceptance of the importance of what might be understood by doctors to be the most 'convenient' sex for the person in question to adopt, given his or her own personal history, desires and the social impact that change of sex in a legal register and in society might entail. In this way, by 1940 we can see the rejection of many of the early nineteenth-century warnings on conceding authority to the patient's account of his/her own feelings as articulated by surgeons such as Marc. In the period 1870–1935 we also see a revision of many of the presumptions of medical authority which inferred that the doctor knew best and was ideally placed to determine the sex of the patient often against his or her will. This, we believe, is a significant development, responding in part to the greater influence of psychological theories of the self, which paved the way towards the creation of the notion of the 'transsexual' in the late 1950s.

This chapter and the next chart these changes by effecting a dual examination of medical texts on hermaphroditism in Spain. The first of these relies upon a 'vertical' analysis of what the major medical disciplines were in which hermaphroditism was considered. One explanation for the differences arising between different or overlapping notions of hermaphroditism, resulting in a non-linear chronology of classificatory systems, relies on the very real divisions in medical thought and practice in Spain. Different disciplines did not necessarily communicate well amongst themselves and certain fields would retain essentially nineteenth-century notions well into the 1930s. The formation of hegemonic opinions in Spanish science was undermined by a lack of centralization, institutionalization and 'paradigm groups', as Glick has illustrated with respect to the growth of the field of endocrinology.[8] In this way, the areas of medico-legal or forensic science, pathological anatomy, gynaecology and sexology could all display differing understandings of hermaphroditism and its significance for their particular discipline or field and could use different systems by which to classify the phenomenon in general or in clinical cases. By means of a vertical analysis, the contribution of these different fields will be analysed with respect to their thought on the question. Many of the sources employed will be medical textbooks or treatises from the various disciplines mentioned above. In addition, however, the way in which this accrued knowledge was exercised, together with the problems it encountered, will be discussed by means of an in-depth examination of a wide range of case studies reproduced in medical journals. While, on the one hand, medical treatises offer a 'state-of-the-art' insight into doctors' understandings, often, on the other hand, considerable doubt and contradictory approaches were discussed and implemented when practitioners were faced with real cases.

In addition to a vertical analysis of these main medical fields, a 'transversal' analysis will look precisely for the interconnections between these areas in order to see the extent of communication between them in their consideration of hermaphroditism. As we shall see, it was the medico-legal profession that focused most urgently on the question of doubtful sex from the early eighteenth century onwards but it took much of its inspiration from the consolidation of pathological anatomy as a discipline in the late eighteenth century, as corpses were 'opened up' in order to analyse what they held in the deep recesses of their tissues and nerves.[9] On the other hand, certain sectors of clinical medicine in Spain even in the late nineteenth century still accepted a largely romantic understanding of the body and its dysfunctions, as will be seen in the case of the eminent José de Letamendi.[10]

Throughout this period, 'hermaphroditism' involved a nexus of concerns including errors of sex, vices of conformation, transvestism, the question of appropriate gender roles, sometimes sexual predilections and the attempt to align anatomy with bodies within traditional frameworks of male- and femaleness. But hermaphroditism in all these accounts waxes and wanes; it is more present in some than in other discussions. Hermaphroditism, in this sense, constitutes a nineteenth- and early twentieth-century device by which all these concerns could come together and be expressed.

The social, political and gendered context of the late nineteenth century

If this process of terminological and diagnostic change was not linear in any simple way, nor was it due merely to 'internal' processes in medical science. Other rationales were at work here, related to the political and social changes wrought by liberalism with respect to the role it reserved for women and the consequent changes in gendered relations, the ordering of the territory in terms of what might be called bio-politics and governmentality,[11] the modernization of society and a cultural challenge to the rigidity of sexual difference particularly by the emerging feminist movement but also by literary and poetic responses to shifting gender possibilities at the *fin de siècle*. While it is easy to over-read the influence of these social, political and cultural changes on doctors' opinions, we can say, without supposing that they drive medical theories, that they do form the backdrop to evolving medical discussions.

In this sense, as part of discussions on hermaphroditism we see the evolution of some very nineteenth-century concerns such as the ordering of subjects according to existing marital law, the refusal to sanction certain criminal acts (such as sodomy, or sex or marriage between individuals of the same sex), the avoidance of disruption in the ranks of the military by defining the (male) sex clearly and the expulsion of any subjects disrupting this category.[12] In addition, it was necessary to make sure that the sexes performed the 'correct' tasks in terms of the labour they undertook as the labouring individual was raised to one of the foundations of subjectivity in the nineteenth century.[13] This swathe of concerns was dedicated more to the ordering of the population according to these criteria by assigning proper roles than it was to elucidating the sexual practices of the individual. However, at the end of the nineteenth century, when taxonomies became less reliable and more individualized, less emphasis was placed in medical accounts on the role of the individual in political, labour, military or marriage terms and more emphasis was placed on the sex and sexual behaviour of the individual as something interesting in themselves.

If the eighteenth century saw the articulation of medical differences between the sexes in terms of the size, weight and structure of the skeleton, the size and function of the brain, and the role to be played by both sexes in society and in reproduction, in the nineteenth century these differences took on an increasingly politicized dimension, illustrating the conflicting legacy of Enlightenment thought and liberal political systems. Although liberalism, as a political articulation of Enlightenment thought, sought to create equality in the legal and social sphere between the sexes and between different kinds of human being, the notion that 'all people are by nature equal was met in conservative quarters with the search for natural differences' with vigour from the early nineteenth century onwards.[14] These 'natural' differences, ascribed to women and non-whites, operated to exclude these groups as political subjects: 'inclusion in the polis rested on notions of *natural* equalities, while exclusion from it rested on notions of *natural* differences'.[15] In the Spanish case, the contradictions of liberalism and the fact that its proponents were mainly men resulted in the recognition of a special sphere for women as different social, political and biological beings. The ways in which Romanticism and the liberal period involved major concessions to class and gender divides, the construction of hierarchies of exclusion and the consolidation of separate spheres for men and women over the nineteenth century have been illustrated by a number of authors, including Bridget Aldaraca, Catherine Jagoe, Jo Labanyi and Susan Kirkpatrick.[16] According to Aldaraca, for nineteenth-century liberal ideology in general and for the followers of Krause in Spain in particular (the vehicle by which much liberal ideology was imported into Spain), there was a marked antagonism in the relation between the public and the domestic spheres whereby the family was seen as the basic cell of an organic whole, state or nation. This contradiction between the public and private and harmonization of the family within the context of a broader political and social project would be resolved by according women a specific civilizing and domestic role within the Christian home.[17] The result of this division of tasks and the faith in the female to make up the cornerstone of the family, society and hence state, was articulated in the notion of the 'ángel del hogar', a concept that emerged from 1850 and was prevalent until the end of the century. This concept 'canonized the woman who accepted her role in the private sphere'.[18] The roots of the 'domestic angel' can be traced to the sixteenth-century text by Juan Luis Vives, *Institutio Foeminae Christianae* (1521) and that of Fray Luis de León, *La perfecta casada* (1583), which posited the woman as a weaker vessel, sinful and morally corruptible. By the nineteenth century, however, Jagoe argues, women came to be associated not with weakness in moral terms but with superior morality, while men were the fallen creatures. This allotted women 'unprecedented spiritual authority in the private sphere'.[19]

Such changes were incorporated into liberal thinking and legislation. Paradoxically, at first sight, liberalism eroded what few rights women had. The 1889 Civil Code reinforced previous legislation on women, whereby wives who disobeyed their husbands received a reprimand, arrest or a fine. Women could no longer dispose freely of property and only they were deemed capable of committing adultery.[20]

The role of medicine in this account cannot be more than subsidiary. Although Jagoe recognizes that marriage manuals reinforced women's role as primarily reproductive, she affirms that psychiatry contradicted the contemporary discourse on femininity by defining the woman as morally deficient.[21] Scientific discourse, Jagoe notes, was caught in a struggle between the models of incommensurable sexual difference and veneration of the female in bourgeois ideology and the older model of patriarchal hierarchy whereby the woman was understood as an inferior copy of man.[22] But, it can be argued, these differences are not necessarily incompatible. While gynaecologists, psychiatrists and sexologists all believed in the need to reinforce and consolidate sex differences, they believed that women were the fount of decency, family stability and sexual modesty and disease, degeneration and immorality *at the same time*. The emphasis on the deep causes of pathological behaviour, well-defined sex differences and the growth of individuality in the late nineteenth century would all contribute to the requirement to find the 'real' sex of 'hermaphrodite' individuals.

Evolving frameworks for the understanding of hermaphroditism, 1840–1910

While the amount of attention that medical doctors paid to the question was not necessarily extensive (medico-legal accounts favoured the subject for reasons already given), we can point to four principal processes with respect to the reception of thought on hermaphroditism in Spain over the years 1840 to 1905. First, we can trace a surge in cases of so-called 'hermaphroditism' from around 1870 to 1890, despite the warnings by Mata and others that the category was inapplicable to human beings. Secondly, as a parallel process, we see how the use of the term hermaphroditism and the meanings attached to it became highly nuanced from the 1840s onwards, being replaced by various different types of pseudo-hermaphroditism from the late 1870s. Thirdly, medical journals are silent on hermaphroditism and related conditions from 1890 up to approximately 1905. Fourth, from 1907 there was another surge in cases reported in the Spanish medical press on pseudo-hermaphroditism and cases of hypospadias (where the urethra opens not at the tip of the penis but anywhere between the glans and the perineum). Conditions such as 'vices of conformation' of the genitalia and hypospadias, previously taken as evidence of hermaphroditism, would undergo a re-evaluation. They would either be slowly disengaged from the category, would support a diagnosis of pseudo-hermaphroditism, or would be taken as states in their own right. Only the most extreme cases of genital ambiguity would be retained in the hermaphrodite category. This expansion of terms overlapped with the entry of ideas on the 'sex glands' from the end of the first decade of the twentieth century onwards.[23] This chapter takes our story up to the early twentieth-century renewed interest in hermaphroditism and the articulation of hormonal explanations for sex.

The medico-legal attempt to determine true sex

As we have seen in Chapters 1 and 2, the emergence of a new professional field, medico-legal or forensic medicine, was, amongst other competencies, to grant doctors the power to pronounce on the sex of the individual in cases of suspected hermaphroditism.[24] The eight-volume *Las leyes ilustradas por las ciencias físicas o tratado de medicina legal y de higiene pública* (Madrid, 1801–3), by F. M. Foderé constitutes one of the first modern treatises of legal medicine to be published in Spain.[25] Up to this time, doctors had been called upon in specific cases of wrongdoing such as witchcraft, poisoning or violence merely to give testimony; now they were called upon to present detailed reports to the courts and their authority generally held sway.[26] In the first third of the nineteenth century, several texts on the subject were published such as P. M. Peiró and J. Rodrigo, *Elementos de medicina y cirugía legal arreglados a la legislación española* (1832), consolidating the discipline into a discernible and powerful body of knowledge.

This trend continued and 1847 saw the publication of Mateo Orfila's *Tratado de medicina legal*. In this text, Orfila made explicit the claims of the new field and situated its role within the modern state. He noted that magistrates sought the views of the doctor and surgeon on many different questions, from public health to the administration of justice, questions relating to the examination of the air people breathed, of the water they drank, and matters pertaining to prisons and epidemics. Defining the 'science' of legal medicine more closely, Orfila explained that it analysed

> el conjunto de los conocimientos físicos y médicos, que pueden ilustrar á los magistrados, al resolver muchas cuestiones concernientes á la administración de la justicia, y dirigir á los legisladores al hacer cierto número de leyes.[27]

the sum of physical and medical knowledge, which can inform magistrates in their resolution of many questions concerning the administration of justice and guide legislators in the drawing up of a certain number of laws.

Legal medicine, following Foderé, was divided by Orfila into two parts: a 'mixed' section that would be applied to civil and criminal issues and to matters of public hygiene, and a second part, which would cover criminal medico-legal practice.[28] Under his section on marriage Orfila discussed the two principal areas with respect to which the medical doctor could be called upon by the authorities to adjudicate.[29] The first of these was to confirm whether the consent given by both parties was indeed valid. The second area involved questions relating to the 'error of the person', that is, if one of the married couple was impotent or belonged to the sex other than 'se habia creido formado parte' (it was thought they belonged).[30] Impotence could imply vices of conformation. In the French legislation, however, the question of impotence was no longer relevant and attention focused on the matter of the error of the person.[31] A note on the ambit of the Spanish legislation, however, clarified that annulment of marriage could take place if the 'miembros necesarios para engendrar' (members necessary for procreation) were absent, if a (male) individual was castrated or if the woman's nature was too narrow to allow for cohabitation with the male.[32]

It was only under a section on the vices of conformation that hermaphroditism was raised; Orfila, as we have seen, dismissed the phenomenon as non-existent amongst human beings. Instead, he argued, vices of conformation in an individual could be the subject of a legal examination since it was necessary to confirm the civil state, that is, sex, of a person or if a declaration on his or her ability to procreate before or after marriage was in question.[33] In this way, Orfila was faithful to Marc in that the latter, on explaining the medico-legal ramifications of cases of hermaphroditism, had noted:

Je n'entrevois que les deux cas suivans où l'hermafrodisme puisse donner lieu à une enquête médico-judiciaire: 1°. lorsqu'il s'agit de rendre l'état civil de son sexe réel, un individu dont les parties génitales présentent un de ces vices de conformation qui font le sujet de notre texte; 2°. lorsqu'il s'agit de statuer sur l'aptitude d'un pareil individu à la procréation et par conséquent au mariage.[34]

I can only foresee two types of case where hermaphrodism could give rise to a medico-legal investigation: (1) where it is necessary to confirm the civil state in accordance with the real sex of an individual whose genitalia present one of the vices of conformation that we are discussing in this text; (2) where it is a case of confirming the aptitude of a similar individual in respect of their ability to procreate and hence marry.

Orfila, however, maintained a certain ambivalence to the categories illustrated by Marc; although he wished to eject the term hermaphroditism from medicine, at the end of his account he fell back on Marc's classifications of six points allowing for the identification of such cases.

A process of disavowal of the category allowed its ambivalent naming and maintenance as a kind of impossible possibility. In a similar way, vices of conformation that previously may have been deemed evidence of hermaphroditism were resignified as precisely that – vices of conformation or organization of the genitalia.[35] But Orfila also allowed for some cases that fell outside of those discussed by him up to this point:

> *Hasta ahora no se ha tratado mas que de los individuos en quienes podria creerse que existen algunos órganos genitales del sexo diferente del suyo, si se examinan de una manera superficial; pero que en realidad, no presenta semejante reunion.*[36]

Up to now we have only discussed those individuals in who one could affirm after an initial examination that there exist some genitalia of the sex different to their own. In reality, however, they do not display such an overlap.

A number of 'monstrous' cases are then cited in which the genitalia of both sexes are mixed. Despite this apparent concession, however, the category of the hermaphrodite is not saved by Orfila – he admits that none of the cases possessed a perfect organization of the genitalia allowing them to 'fecundar y ser fecundado' (impregnate and be impregnated).[37]

By the time we arrive at the 1850s the pedigree provided by legal medicine with respect to the analysis and determination of sex was already well established as one of the areas of its competency. Despite this, however, we have noted a number of ambiguities and ambivalences on the way, although hermaphroditism was deemed a (practically) impossible category in human beings. These kinds of equivocal accounts would remain in studies of hermaphroditism throughout the period studied and are seen in the work of our next figure, Pedro Mata i Fontanet, who took Spain's first chair in legal medicine at Madrid in 1844.[38]

The most extensive account of sexual abnormalities, including the emerging category of sexual inversion and the question of sex identification from this early period, can be found in Mata's work. His *Tratado de medicina y cirugía legal* enjoyed considerable success with a total of six editions with an expanding number of volumes being published between 1844 and 1903. Mata had gone to Paris to work under Orfila in 1837, returning to Barcelona in 1840. While there are, as one would expect, a large number of continuities between the successive editions of Mata's treatise, there are also some substantial differences. An important one, in terms of medico-legal practice, was the assumption of new diagnostic frameworks for identifying sexual categories and misdemeanours derived from the work of the French doctor Ambroise Tardieu.

In the same way as the arrival of the thought of Tardieu revolutionized the ways in which pederasty and sodomy were considered in Spain in the 1860s, for example in the writings of Mata, a similar process occurred with hermaphroditism. In the editions of the *Tratado* published in 1844, 1857 and 1874 the sophistication with which impotence, hermaphroditism and the marriage question were treated varies considerably. In Mata's 1844 *Vade Mecum de medicina y cirugía legal*,[39] the reality of the hermaphrodite was denied in words that recall those of Orfila – such a trait can be understood as 'aquella disposicion viciosa de las partes genitales por la que el individuo parece ser de un sexo á que realmente no pertenece, ó no se puede determinar cual sea su verdadero sexo' (that vicious disposition of the genital parts that makes the individual appear to be of the sex to which he does not belong or for whom one cannot determine their true sex). Any 'malas conformaciones del aparato génito-urinario' (irregular conformation of the genito-urinary apparatus) can be divided into three types: masculine, feminine and neutral or epicene, following Marc's system from 1817.[40] A number of case studies common to European discussions followed, including Maria Margarita and Maria Lefort. The characteristics of hermaphroditism were detailed in this 1844 volume comprising in the masculine case: the appearance, propensities, timbre of voice, habits of a man; occasional development of breasts and lack of inclination towards the female sex; the scrotum divided into two clear parts with (semi-) hidden testicles; and with the penis small, imperforate, or with hypospadias. In the female hermaphrodite, there would be general masculine forms in terms of voice, body hair and muscular strength in some cases and in others the feminine would preside. The clitoris would be large, there would be no vulva and the opening of the urethra would be at the base of the clitoris leading to the vagina. Urine and menstrual flows would be expelled from here. In the case of the neutral or epicene, the individual would display characteristics of both male and female, in terms of genitalia and other parts and aspects.[41] It is interesting to note how in the male hermaphrodite no female countenance is recorded. In addition, while the question of sexual leanings is mentioned in the male case, no such characteristic is deemed important in the female case, showing how the female was still considered incapable of significant autonomous sexual desire or displayed uncontrollable sexual inclinations.[42] In any case, for an account that denied the existence of the real hermaphrodite, these lists constituted an extensive typology.

The main question surrounding the question of hermaphroditism was the individual's ability to reproduce, following the second of Marc's criteria for examination of hermaphrodites. Mata stated that, in the male case, if there were testicles, if the urethra communicated with the bladder and the sperm vesicles and if this opening could deposit sperm naturally or artificially in the vagina, the person was deemed 'potent'. In the female case, a vulva, vagina, uterus and the possibility of allowing the penis to enter and to accept sperm, were all necessary to declare the woman 'potent'. In the case of the epicene, he or she could be declared potent according to the degree of development of the organs. It is important to note that this was the overriding factor for the medical doctor: Mata states 'Si el hermafrodita es neutro ó epiceno, sin que se pueda determinar á qué sexo pertenece, debe ser declarado impotente ó potente, segun el desarrollo ó conformidad de sus órganos' (If the hermaphrodite is neutral or epicene and one cannot determine the sex to which he or she belongs, he should be declared either impotent or potent, in accordance with the development or conformation of their organs).

This is where the 1844 edition stops. The 1857 edition of Mata's work more or less reproduces this classification but then continues into other territory. What follows the basic description of the categories is a long analysis of the minutiae of the in/ability to fertilize or be fertilized. Mata affirms that even in cases of hermaphroditism, following the outline above, no difficulty should be encountered in determining this question. However, other cases of a more difficult nature were often encountered. One difficulty, 'anaphrodisia',[43] nervous impotence or the lack of testicles in men, was not necessarily easy to detect. Was the man able to achieve both erection and ejaculation? Different scenarios were created by the accidental loss of the testes and the state of cryptorchidism (where the testicles have not descended). In the latter case, the male possesses, in most cases, all the characteristics of virility. In some cases, feminine countenance is present but not to the degree shared by those who 'carecen de dichas glándulas' (lack those glands).[44] In the case of the eunuch, female characteristics predominate. In those who have never had testicles, 'los sentimientos duermen, las pasiones no se revelan en ellos, no hay ni instinto, ni idea de placer venéreo' (sensation dies away, the passions are not present, there is no instinct or concept of venereal pleasure).[45] The lack of potency in such individuals is clear.

What should be done in the case of the cryptorchid individual? The potency of the individual either way could be declared impossible to determine, or an analysis of the sperm, with the aid of a microscope, could be effected. The moral problems surrounding the collection of sperm are discussed at length. Mata analyses these questions with the aid of the example of a recently married couple he was called to adjudicate as potent or otherwise with Drs Ulibarri and Cuadra, the report on whom is reproduced in this 1857 edition. What follows for the couple 'D.N.N.' and 'D^a.N.N.' is an intricate and intimate examination of sperm, in the first case, and the hymen and vagina, amongst all other bodily elements, in the second. Both are declared to be potent. Hermaphroditism and this case are, as we would expect, included under the section dedicated to the determination of potency in human beings. What is at stake in the example of D.N.N. and D^a.N.N. is not whether one or the other is a 'hermaphrodite' but whether they can cohabit in order to reproduce.[46]

A secondary question, discussed on a single page of the 1857 edition, is the matter of error of sex in marriage. Mata here was not referring to what he called an error in the 'identidad del sugeto' (identity of the subject), by which one found out one had married Pedro Méndez rather than Juan Álvarez. This was a question of identity.[47] An error of sex, or error of person (Mata uses both terms) would occur if a man married a person he thought was a woman or vice versa. This followed Marc's first criteria for examining hermaphroditism. In order for this 'confusion' to happen, hermaphroditism must be present. Mata argues that cases such as Lefort and Margarita confirmed that there would have been an error of sex. In such cases, the marriage should be annulled. If a male hermaphrodite is married to a man, the marriage is annulled. If a female hermaphrodite is married to a woman, the marriage is in turn annulled.[48]

Mata's fifth edition of the *Tratado* of 1874 included the section on hermaphroditism, following previous medico-legal conventions, in the section which was dedicated to 'cuestiones que versan sobre el estado y funciones de los órganos sexuales ó su producto' (questions that deal with the state and functions of the sexual organs or their product).[49] In this section, there were seven chapters. The first centred on 'cuestiones relativas al matrimonio' (questions relevant to matrimony), the second on incontinence and crimes against morality, the third on pregnancy, the fourth on birth, the fifth on abortion, the sixth on premature and late births and the last on 'superfetación', the overdevelopment of the foetus. Hermaphroditism is discussed after the section on impotence, as had been the case with other medico-legal texts discussed above. The fifth edition contained much of the same material as the previous editions, word by word, with similar stories from abroad, the same system of classification (with the masculine, feminine and epicene hermaphrodite), similar comments on cryptorchid males, the means of determining potency and an expanded number of case studies.[50] The now usual foreign case studies were repeated and in the case of Justine Jumas Mata made reference to Carcassonne, Tardieu and Legrand du Saulle as his sources.[51] Notably, he discussed the case of Alexina B., referred to extensively by Tardieu, from whom he probably derived his source material.[52] The 1874 edition saw more Spanish cases come to light in which there was some dispute over the sex of the person or their potency.

The first of these cases from 1846 discussed an army recruit interned in the Hospital Militar de Madrid.[53] The second of these cases, originally from 1847, related the experience of Mata's colleague Dr Carrasco. Carrasco showed Mata the genitalia, preserved in alcohol, of a girl who was identified as a boy on birth as a result of the 'enormous' clitoris, 'like a penis', which the child possessed. The autopsy of the unfortunate girl showed internal organs corresponding to the female sex. In a twist of medical labelling the 'boy' was renamed a 'female hermaphrodite', that is, an apparent boy who was in reality a girl, after the medical examination of her internal organs.[54] It is worthy of note that, despite all the increasingly sophisticated taxonomies followed by Mata, drawing on Carcassonne, Tardieu, Marc and Worbe (note that Mata called the female hermaphrodite a 'gynandryne', a term absent from previous editions), old medieval concepts of the large clitoris as a sign of hermaphrodites were still in use. Such a sign had come under increasing scrutiny in the eighteenth century.

A third case in this edition is worthy of note. Similar to the case of Dª. N. N., this certification of sex from 1860 focused on 'R.V. é I.' This woman was the subject of two articles in the important *España Médica* in 1860 and 1861. It is interesting to note that Mata titled his discussion of R.V. é I. as a 'certificación de sexo de una persona' while his colleague Alba y López described it as a case of hermaphroditism.[55] The significance of the case is that it enables us to see once more how we are faced with differing medical diagnoses and, more importantly, different (although related) driving motives behind the examination of the individual concerned. An analysis of both accounts will illustrate this process.

The report by Mata is drier, more descriptive of the body of the woman concerned and more focused on the requirement to fix the sex of the individual and to determine whether she was potent. The report by Alba y López is, at once, more technical and more grandiose in its claims. In cases such as these usually it is difficult to determine precisely why the individual ended up in the clinic, as their voice is often obliterated by that of the physician, but here we learn from Alba that the 24-year-old, single, Madrid-born, servant, came to the clinic of Dr Ulibarri complaining of hip and body pains. The case is of great interest to all concerned. Mata defines it as 'un caso curioso para la ciencia' (an interesting case for science); Alba refers to 'la rareza del caso' (the unusual nature of the case) and to 'lo estraño del caso' (the strangeness of the case).[56] A certain hyperbole surrounds Alba's account. The patient has to be won over before she will allow the assembled doctors to examine her. The power of 'un gran número de catedráticos y profesores de medicina y cirugía' (a large number of professors and doctors in medicine and surgery), in Mata's words,[57] must have been overwhelming for a 'servant girl'. Indeed, her resistance was overcome by 'la persuasiva voz de nuestro digno catedrático' (the persuasive tones of our renowned professor, Dr Ulibarri), who explained the use such an examination would perform for humanity. There was no need to worry, in any case, as the clinic was 'el templo del silencio' (the temple of silence), inferring that any discussions would be kept private. Convinced by such 'solid reasons', 'no se negó la enferma, y se presto gustosa á satisfacer la necesidad de la ciencia' (the patient did not refuse and she disposed herself to satisfy the needs of science).[58] This kind of reasoning prevailed throughout the report and was repeated to justify the examination and the impending intervention that the doctors would apply.[59]

In the visual accounts of R.V. é I. the physical characteristics of the woman were deemed more or less normal save a number of gendered elements that implied, for the doctors, masculinity. Alba speaks of 'laringe y voz masculinas' (male larynx or voice), 'carece de glándulas mamarias, y presenta solo las mamilas masculinas rodeadas de vello' (she lacks mammary glands and only presents male nipples surrounded by hair); Mata records a general masculine aspect, with a 'mirada enérgica' (energetic gaze), 'labios, carrillos y barba provistos de pelo, los cuales se afeita' (upper lip, cheeks and chin populated by hair, which she shaves), 'voz fuerte, sonora, varonil' (strong resounding masculine voice). What intrigue the doctors, however, are certain irregularities in the genitalia of the woman, about which there seem to be different diagnoses. For Mata, R.V. é I. may well at first sight 'pasar por un hermafrodita y dar lugar á dudar de la realidad de su sexo' (pass as a hermaphrodite and give rise to doubt over her sex), but once properly examined, she is clearly a 'true woman'. The only aspect that could cause confusion is 'un cuerpo cilíndrico de unas tres pulgadas de longitud ... enteramente parecido al exterior á un pene imperforado, con su glande y su prepucio' (a cylindrical body of some 3 inches in length ... completely similar externally to an imperforate penis with a glans and prepuce). This 'body' is capable of erection, according to the patient; there is also a 'rudimentary clitoris' and labia beneath this 'especie de pene' (species of penis). The vagina allows for the introduction of an index finger and there is evidence of a womb, although underdeveloped, given the lack of, but not complete absence of, menstruation.

Mata's account is somewhat different from that of his colleague Alba y López. Alba y López records 'un pene de unas dos pulgadas de longitud' (a penis of approximately 2 inches in length), he suspects a hypospadias of the third degree, but agrees that the womb exists 'in a rudimentary state'.[60] Dr Ulibarri prescribes certain medicaments to alleviate the condition and the resultant pain. Alba continually refers to the 'hermaphrodite', and the problem appears to be the 'penis-like' form discovered by the doctors. An operation is suggested to 'correct' this, something that the patient wished for once she had understood 'la deformidad que aquejaba' (the deformity she suffered from). She began to put pressure on Dr Ulibarri to operate.

Alba y López notes that it was here that differences of opinion came to light. Some of those present advocated the operation while others deemed it unnecessary. Alba, acknowledging that R.V. é I. constituted a case of feminine hermaphroditism (of masculine appearance but in reality a woman), believed that an operation was necessary; it was not, he stated, 'una operación de complacencia' (an operation of convenience) as the 'penis' formed an obstacle for normal coitus. Penile amputations, he considered, took place with great success. What followed was an account of the successful operation (Mata also records that the operation took place) and a justification of the procedure as evidence of the value of science in its alleviation of human suffering.

For Alba, then, R.V. é I. constituted a case of female hermaphroditism, which science could correct by means of an operation that would substantially reduce the offending item, a 'penis' that was imperforate and in an individual who possessed a womb and menstruated. In his account there was no mention of the sexual desires of the patient, unlike in France and Britain where, according to Dreger's analysis, it was believed that quizzing people of doubtful sex about their sexual organs and desires would elucidate the nature of their real sex – testicles meant desire for women and ovaries desires for men. Some doctors, like Pozzi, suggested that sexual predilections had no importance in these cases but few practitioners heeded their warnings.[61] In part it would appear that this responded to a desire to eliminate the recognition of any actual homosexuality. But it also responded to the mistrust with which doctors held patients' own analysis, following Marc's warnings.

Mata, as we have said, had already rejected the finding of hermaphroditism of any variety in R.V. é I. While, as his *Tratado* argued, it was necessary to detect any suspected hermaphroditism in individuals this was in order to certify sex and to determine the potency of the individual. His account of R.V. é I. followed these concerns to the letter. His 'Certificación sobre el sexo' proclaimed that R.V. é I. possessed all the necessary organs for copulation and for the reception of the 'licor prolífico en vaso idóneo' (prolific liquid in the correct vessel). This enabled him to confirm that R. was 'verdaderamente una persona del sexo femenino y potente; que como tal puede contraer matrimonio' (truly a person of the female sex and potent; as such she can marry). On the question of children, however, Mata observed that it was not possible to 'asegurarse que tenga hijos, como no se puede asegurar de cualquier otra, por bien organizada que esté' (affirm that she can have children, just as it cannot be affirmed in another woman, however well organized she may be). Mata was more interested in assuring that this woman and others were 'well organized' or properly disposed to facilitate such tasks, from a now classic medico-legal stance.

Alba (and perhaps Ulibarri), from the perspective of pathological anatomy, was more interested in making sure that deviant bodies were properly managed and that sex, gender and body were restored to harmony, even though (heterosexual) activity was clearly a priority (the penis-like form was deemed an impediment for coitus). Alba also set more store by the category hermaphroditism as a specific quality in its own right. Despite these differences, which responded to distinct disciplinary concerns, what united all doctors was an emphasis on a visual and physical examination of the patient, where the genitalia of the person in question, the presence or otherwise of a womb, secondary characteristics such as hair, and tertiary characteristics such as his or her comportment were analysed in order to produce 'true sex'. It is striking that no ovaries, no suspected testicles and neither the sexual behaviour of the individual nor his/her sexual desires or 'orientation' were mentioned in either account.

The disruption of gendered relations

The desire to restore harmony in cases of dubious sex meant that the disruption of gendered relations was at the heart of a number of cases recorded from the 1860s to the early twentieth century. These concerns were often expressed in terms of 'error of sex' or 'error of person', along the lines that Mata had discussed. Hermaphroditism was more present in some of these cases than in others; in some, it formed a kind of lurking suspicion rather than a concrete diagnosis.

Cheryl Chase has emphasized the very real implications for persons caught up in this medico-legal network of power: 'The insistence on two clearly distinguished sexes has calamitous personal consequences for the many individuals who arrive in the world with sexual anatomy that fails to be easily distinguished as male or female.'[62] This could not have been truer for the first case that we consider. In the mid-1860s José Pablo Pérez and Carlos Cherizola described the case of María de los Reyes Carrasco y Huelva, an individual whose ambiguous sex and misfortunes took her from her native Puebla de Guzmán (Huelva province) to Portugal, Malta and Cadiz.[63]

The two authors of the account, as faculty attached to the hospital in Huelva, understood the deviant acts of their subject as a consequence of the deviant anatomy she displayed: her 'historia está llena de estrañas aventuras hijas todas de la irregularidad que ofrece su aparato generador' (story is full of strange adventures, all the product of the irregularities offered by her reproductive apparatus).[64] Reyes Carrasco had clearly made an impact on the locality as she was regarded by local public opinion as a hermaphrodite; the authors discussed her case from a 'physical and moral' perspective. A description of María follows and a certain number of characteristics breaking with her female gender are noted: she can lift heavier weights than practically any other individual; she wears her hair short; has a guttural voice, and 'sus inclinaciones son enteramente varoniles, prefiriendo el trato y costumbres del sexo opuesto al suyo' (her inclinations are supremely virile, preferring contact and custom with the opposite sex).[65] While these 'inclinations' do not convey any notion of sexual preference, the doctors' account of them displays a clear regard for deep sexual differences between men and women. Further physical differences seem to differentiate her from the standard woman. She lacks pronounced breasts, her clitoris 'se asemeja á un pene de medianas dimensiones' (is similar to a penis of average dimensions), although it is imperforate, and she has a narrow vulva. Any menstruation was short-lived. There was some doubt over her sexual identity as a child but when the date arrived for her to dress in a manner 'appropriate to her sex', a local doctor advised the use of female attire. However, subject to the taunting of her peers, who wished to see her 'extended' clitoris, she pleaded with her parents to be allowed to change dress. They consented and she worked as a farm labourer and later mule driver, taking the name Manuel.

Subsequently, she went to work as a miner in Portugal but, recognized as a woman by a neighbour who also worked there, got into a fight with this neighbour and stabbed her tormentor. She fled, tried to go to London by boat, ended up in Malta after a dispute on board, was arrested as a suspected army deserter, was packed off to Barcelona but escaped to Gerona. Here she was detained and declared her true name, nationality and sex (omitting the details about the knife fight in Portugal). The latter event was discovered, however, and she was subject to a medical examination and 'considerada como un tipo curioso' (considered as an interesting individual). She was sentenced to seven months in prison because she did not possess any documentation certifying her name and profession, spent time in jail in Barcelona, and returned afterwards to Cadiz. On returning, her previous deed in Portugal was recounted and the authorities of this country aimed to try her for murder. This development caused a severe bout of typhus and María/Manuel was interned in hospital for six months.

Reyes denotes an individual at odds with late nineteenth-century Spain. She is a figure operating between two worlds. On the one hand, as in medieval times and up to the seventeenth century, any person of doubtful sex would be assigned the dress of the sex they approximated to most or they wished to settle as (without being able to reverse that decision). The town of Huelva in the late 1840s when María/Manuel was a child seems to have accepted this without too much difficulty, forming a striking difference with what was to occur to similar persons at the end of the century. On the other hand, she became an individual without identity in an age when the possession of such a mark with the requisite documentation was important. In addition, her wanderings and misdemeanours were understood as deriving from her unusual anatomy: 'lo estraño de este caso se halla mas en la vida de relacion que en la deformidad fisica' (what is strange about this case has more to do with social relations than any physical deformity). Although her classification had not been mentioned before in the article, except in its title, the authors finished their report by summarizing that her physical 'deformity' was 'la forma mas ordinaria de lo que ha dado en llamarse HERMAFRODISMO' (the most common form of what has been called HERMAPHRODISM). The epithet of hermaphrodism accounted for all her deviancies to date.

The relationship between gendered social position, anatomy and the need to account for 'errors of sex' was another conjunction of concerns that drove what legal doctors such as Orfila and Mata had to say in the context of hermaphroditism. The 1898 volume written by Legrand du Saulle, Georges Berryer and Gabriel Pouchet analysed the medico-legal position with respect to marriage and its annulment from the basis of the French, Spanish and other legislations, with Teodoro Yáñez's and Carlos Núñez Granés' commentaries on Spanish law.[66]

Under the section on marriage with respect to the French legislation, the three principal authors of this volume discuss errors of person, which are nearly always, they argue more explicitly than Orfila did in 1847, errors of sex. In this light, they relate the case of Alexina B. as an example of an error of sex.[67] In a short section afterwards, the authors elaborated the framework to follow in cases of doubtful sex. Their model relied principally on Marc, St-Hilaire, Briand and Chaudé. There would be three stages in the determination of sex. First, the doctor should observe over a long period of time the gestures and habits of the individual, 'teniendo siempre cuidado de no confundir los hábitos que pueden resultar de la posición social con las propensiones innatas ó resultantes de la constitución orgánica' (taking care not to confuse those habits that may result from social position with innate propensities or those resulting from organic constitution).[68] Such a procedure is faithful to Marc's sixth observation.[69]

Secondly, it was necessary 'Determinar, después de la inspección de toda la superficie del cuerpo, cuál es el sexo cuyos caracteres parecen predominar' (to determine after due inspection of the whole of the body's surface, which is the sex whose characteristics seem to be predominant). Third, doctors should examine

> con mucho cuidado las partes exteriores de la generación, y sondar todo lo que sea posible, sin excitar un vivo dolor, todas las aperturas que se presenten, á fin de conocer su extensión y su dirección, descubriendo así los vicios de conformación que ocultan el verdadero sexo.[70]

with much care the external organs of generation and investigate as far as is possible, without causing any pain, all orifices in order to find out their depth and direction. In this way, the vices of conformation that hide the real sex will be discovered.

The diagnosis remained faithful to the main postulates of Marc's system with an emphasis on visual and tactile strategies. The authors also argued that it was necessary to determine whether there was evidence of menstrual blood, taken as sufficient evidence to declare the subject a woman. Here there operated a presumption – that a woman who menstruated was indeed 'potent'. The invasive examination of 'the apertures' would most likely be effected in women. The examination of sperm was not mentioned as a procedure.

The above discussions related to the question of error of person and/or sex. In a subsequent section, Legrand du Saulle, Berryer and Pouchet outlined their thoughts on hermaphroditism per se. Here, they reiterated what they announced to be St-Hilaire's construction of three categories of hermaphrodite: the exact positive androgyne, a category that contained true males and females but whose 'conformación tan viciosa' (so vicious a conformation) made sex determination impossible; the approximate positive androgyne, whose sexual apparatus is neither completely male or female, but in whom one sex predominates, allowing in some cases for the appropriate sexual functioning with respect to that sex; and the neutral negative hermaphrodite in whom the sex is indeterminate, arrested in its development and placed between the two sexes (also referred to as the 'mixed hermaphrodite').[71] In reality, however, this classification did not correspond to that of St-Hilaire.[72]

What the trio of doctors held to be St-Hilaire's system, however, was not deemed satisfactory because it was thought to be overcomplex. Instead, 'El médico-legista escrupuloso no puede conformarse con esta clasificación. Para él no hay más que dos categorías: la de sexo definido y la de sexo que no puede ser definido' (the scrupulous medico-legal doctor cannot be satisfied with this classification. For him there are but two categories: defined sex and sex that cannot be defined). This apparent simplification of the question would aid the doctor in determining the legality of a particular marriage: 'El matrimonio de todo hermafrodita de sexo más ó menos dudoso *es nulo*, no por error en la persona, sino por identidad de sexo entre los dos esposos' (the marriage of any hermaphrodite of more or less doubtful sex is to be declared *null and void*, not because of error of person, but because of the sex identity of the two partners).[73] Once again, the important driving criteria relied on the avoidance of marriage between persons of the same sex and the confirmation of potency.

The framework employed by these doctors was to become more eclectic as the section advanced. In cases, admittedly extremely rare, where an individual possessed the attributes of both sexes, following Tardieu and Brouardel, there would also be 'identidad de sexo' – sameness of sex – and the marriage would be annulled because the real sex of the person could not be determined incontestably. The law and the judge demanded 'no probabilidades, sino certidumbre; no se trata de una cuestión de impotencia posible ó probable, sino de una cuestión de identidad de sexo' (not probabilities, but certainty; it is not a question of possible or probable impotency, but a question of sex identity).[74] In one sense, Legrand du Saulle, Berryer and Pouchet attempted to simplify and to clarify cases of doubtful sex by reducing the number of possible categories – from three under St-Hilaire (as they understood him) down to a sensible binary of either a defined sex or an undefined sex – but on the other hand their admission that some cases were indeterminate hardly underpins anything like a definitive system of classification. The three French authors found themselves in between the fading old paradigms of Marc and St-Hilaire and the new 'certainties' occasioned by the new paradigm of Klebs, discussed below.

Towards individualized diagnosis

The *Tratado de medicina legal* by Legrand du Saulle, Berryer and Pouchet was published in Spain in and around the time of the recording of a steady number of cases of doubtful sex or proclaimed hermaphroditism. The need to focus on individual cases rather than any overarching system of classification evidently gained prevalence from the late 1870s onwards. The 1888 case from Canjayar (Almeria) illustrates this shift. In the *Revista de Medicina y Cirugía Prácticas*, J. M. M. (probably Dr Juan Manuel Mariani, of the Hospital Princesa, Madrid) discussed the report by Manuel Sánchez y Sánchez on the subject of a deceased woman ('ó que usaba por lo menos el traje de este sexo' (or who at least used the clothes of this sex)), aged 54, of a strong and vigorous constitution, who worked in the fields performing 'trabajos rudos propios del sexo masculino' (manual tasks characteristic of the male sex).[75] During the autopsy, the external aspect of the body was found to be completely masculine, according to the report, and the penis was of normal size and complete. Between two folds of skin which ended at the perineum, there appeared to be the blind entrance to a vagina. In the skin folds, however, testicles were found. Where the two folds came together was an orifice corresponding to the urethra. Seminal vesicles were present. While under other circumstances this woman may have been labelled a hypospadiac male, here there is no diagnosis either of hermaphroditism, of 'true sex' or of anything else. What is most alarming, apparently, to the investigators is the gender role switch in respect of the role and dress of the person. The 'vagina' merely seems to be a concession to possible femaleness. The individual's upbringing was not recorded.

Cases such as this were analysed as classification systems were becoming more complex and increasingly individualized. In fact, despite the apparently confident air with which Mata and others classified hermaphrodites, and continued to do so well into the early twentieth century, the system was being placed increasingly under strain from a number of perspectives. Already by the date of publication of Legrand du Saulle, Berryer and Pouchet's volume three related processes were in full swing. First, we see by the early 1870s a creeping realization by doctors that cases were becoming more difficult to define and that an increasing number of exceptions to the rules are alluded to as cases go beyond the limits of the existing classificatory systems accepted by legal medicine. For example, in 1872, the volume *Manual completo de medicina legal y toxicología* written by three prominent foreign authors, J. Briand, J. Bouis and J. L. Casper, question Marc's framework but do not suggest an alternative.[76]

Secondly, this process of crisis of the classificatory systems may have cautioned doctors against pronouncing too quickly on the sex of the individual in question. Legrand du Saulle, Berryer and Pouchet warned against premature or overzealous certifications of sex of the new-born. Here, they advocated caution when the genitalia were ambiguous. In order to support their stance they cited a case referred to by Briand and Chaudé from 1816 in which it was decided to change the birth certificate of an individual who was declared to be female although she in fact belonged to the male sex. The authorities at the time allowed the parents to decide the sex of the boy. Could not the decision wait, they argued, until later in life precisely to avoid confusion and harm?[77]

The three French doctors seem to have captured the moment for there is evidence to suggest that some Spanish doctors were entertaining doubts as to when to designate the true sex of the individuals brought to their attention. Two examples will suffice. The first was recorded in 1879. Dr Peset Cervera recalled how his colleague

Dr José Douday of the Valencia civil hospital invited him to examine a baby girl of three months who had been baptized with the name Consuelo S.[78] In this case, it was the parents who wanted to change the girl's sex on the civil register 'en vista de un error de sexo' (in view of an error of sex). Dr Cervera noted that the baby did possess a 'very well formed penis' of approximately 3cm but without a fully formed urethra. A testis was located in the left-hand scrotal sac alone. The vulva measured 1cm but the extension of the vagina remained unrecorded because of the 'especie de himen' (species of hymen) in its way. The opinion of the doctor was such that 'No puede ménos de considerarse á *la niña* dotada del sexo contrario, si bien con un hipospadias completo' (One can only consider that 'the girl' is made up of the opposite sex, although with complete hypospadias). The medical authority is equivocal, however: it does not concede the parents' demands but, given the impossibility of exploring the vagina, 'creí oportuno aconsejar á los padres que permanecieran *in statu quo* sobre la rectificacion de la partida de bautismo hasta que sea llegada la edad de las pasiones' (I believed it opportune to advise the parents to leave things as they were in respect of any changes to the birth certificate until the age when the passions emerge).

The equivocal genitalia in this case seem to carry less authoritative weight than what will eventually be the sexual orientation of the subject, which also presides over the presence of the male 'gonad'. Rather than hermaphroditism being proclaimed, hypospadias is the diagnosis. Furthermore, in tune with what seemed to prevail in British and French medicine, sexual orientation, when fundamental doubt existed, would be the marker of true sex, despite the warnings by influential surgeons such as Pozzi.[79] In this case, the gonads still do not rule the day but neither, apparently, do the genitalia in any definitive manner.

A second example of the decay of classification systems is shown by a case that first appeared in *El Siglo Médico* and which was commented upon in the *Revista de Medicina y Cirugía Prácticas*.[80] This case of 'apparent hermaphrodism' from Alcira (Valencia) focused on a new-born who had been assigned the female sex. There was some doubt at the time of her birth and Dr Alejandro Settier, who examined the case, found 'dos cuerpos de la consistencia, tamaño y forma de los testículos' (two bodies of the consistency, size and form of testicles). The child possessed a clitoris, however, at the base of which the urethra was positioned. This situation led Dr Settier to advise waiting until the onset of puberty in order to classify the case 'por el prominente desarrollo de unos ú otros órganos' (in light of the predominant development of one or other set of organs). In contrast to the previous case, here it would be the genitalia and other organs that would speak the truth of the individual. Sexual orientation, in any explicit manner at least, would not define the patient's sex. Of course, it is possible that, given the lack of discursive space for homosexuality, sexual orientation would be seen equally as a marker of sex as a heterosexual interest would be seen to proceed *naturally* from the genitalia. That is to say, sexual preference was seen to derive directly from the male or female genitalia the individual possessed.

'El campo del hermafrodismo se va estrechando cada vez más' (Pozzi)[81]

A third characteristic of the crisis of the 1870s was, as we have in fact seen in these two last examples, a reconsideration of the boundaries of hermaphroditism, whereby a dual process takes place. In the first instance, the use of terms such as 'apparent' and 'pseudo' are increasingly adopted as prefixes for hermaphroditism. Although from the mid-nineteenth century up to the 1870s Mata and other medico-legal figures had stated that hermaphroditism did not really exist in humans they still classified doubtful cases under the heading of male, female or neutral (epicene) hermaphroditism. Even though Marc had spoken of *apparent* hermaphroditism, it was only around the time of the publication of the sixth edition of Mata's *Tratado* (1874) that other classification systems came on stream which allowed for 'pseudo-hermaphroditism'. The significance of 'hermaphroditism' becomes increasingly difficult to maintain intact as the 1870s wear on and as more and more cases of 'pseudo-hermaphroditism' appear. In the second instance, more cases are classified under the heading of hypospadias as opposed or in contradistinction to hermaphroditism, possibly as operation techniques were improved to rectify the condition.[82] Even though Mata back in 1844 in his *Vade Mecum* had classified hypospadias as an element enabling the classification of hermaphroditism, what takes place from the 1870s onwards is the partial decoupling of the two concepts.[83] This was no definitive replacement, as we shall see, but rather a trend. We have already seen the case in Valencia in 1879 presided over by Dr José Douday. Rather than hermaphroditism, Douday diagnosed hypospadias. Again instead of hermaphroditism, Dr Settier's 1888 analysis ruled 'apparent' hermaphroditism.

This bridging of hermaphroditism and hypospadias is illustrated by another case from 1877.[84] The language expressing the findings in this case from Alicante was more definitive than in many other cases. The subject was declared right from the start of the two-page report to be a 7-year-old boy, about whom some doubt had been expressed on birth. The parents wished for a further confirmation of sex and the doctor in charge, Esquerdo, found, in this order, two 'large labia' that occupied the place of the scrotum and a penis of one and a half centimetres without a urethra. Only one normal testicle was found to be located in the left labia. There was, however, a cavity connected to the bladder which could be probed to a limited extent. While this case could have deemed one of 'male hermaphroditism', that is, denoting a person of male sex but with female elements or appearance, or even epicene hermaphroditism, the following was stated: 'Dedúcese de todo esto, que se trataba de un caso de hermafrodismo ó simulacion de tal, existiendo un hipospadias que correspondia al límite posterior de la porcion esponjosa de la uretra' (From all this it is deduced that here is a case of hermaphrodism or simulation of the same; there exists a hypospadias that corresponds to the lower limit of the spongy part of the urethra).[85] An account of the operation to be performed to correct this was given. Other doctors present argued that given the small size of the penis, corrective surgery would have to wait until the child was older.

As we have said, however, diagnoses of hypospadias (and suggested corrective surgery) did not replace the notion of hermaphroditism swiftly or definitively. Two cases from 1880 would in fact suggest the opposite. One from Barcelona, for example, was declared *not* to represent hypospadias but 'asymmetrical hermaphroditism'.[86] Another from Barcelona recounted the story of 'N.N.', a 28-year-old single male agricultural labourer.[87] The fact that the individual possessed sexual organs of both sexes 'abonan el calificativo de hermafrodita que damos a este individuo' (confirms the classification of hermaphrodite that we have given to this individual). A description of the sexual organs followed and these were deemed to offer evidence of both sexes but in sufficiently separate measure for the male and female elements to be studied independently. The male genitalia were positioned on the right-hand side, consisting of a single larger than normal testicle, a smaller than usual penis (of 3cm) but of greater size once erect. No prostate was discovered by means of a rectal examination and no evidence of another testicle was found, even in an undescended state.

The female organs occupied the left-hand side and a vagina of some 4cm was located. A major and a minor labium were found on the left but not on the right. It was not possible to locate the existence or lack of existence of a uterus. There was no clitoris, it was affirmed, but with some doubt: 'falta el clitoris que, dada la posición relativa de los órganos, casi podríamos considerarlo confundido con el pene' (the clitoris is lacking; given the relative position of the organs we could almost consider it as a penis). The possibility of hypospadias was entertained but rejected: 'El meato urinario se observa en el glande, de modo que no hay hipospadias tan frecuentes en casos de esta naturaleza' (the urinary tract can be observed in the glans, so hypospadias is not so frequent in cases of this nature).

Given these data, the doctors proclaimed the subject to be a *neutral* (hermaphrodite). Although the system of classification is not stated, such an epithet derived from Marc's early nineteenth-century criteria and from St-Hilaire's first class, third order 'neuter hermaphroditism', in which the sexual apparatus presented 'some intermediary conditions between those of the male and those of the female, and being really of no sex'.[88] It is possible that the case corresponded more closely to St-Hilaire's first class, fourth order hermaphrodite: 'Mixed hermaphrodism: sexual apparatus in part male and in part female'.[89] This kind of case would eventually be further reformulated to represent what Simpson had named 'lateral true hermaphroditism' in 1839,[90] a designation retained by Klebs.[91]

The complexities of the case, however, meant that even the designation of 'neutral' was soon nuanced. As the account becomes increasingly detailed as part of a process of social and medical individualization more details are made explicit. That the breasts were not developed and there was a fair amount of body and facial hair suggested belonging to the male sex. As the account proceeds, the genitalia assume relatively less importance (the 'gonads' as such are not mentioned) and a variety of characteristics outlined by Marc are referred to. Moral details and gendered qualities were also relevant in the assessment and these appeared to be mixed:

podemos ver en su carácter y cualidades morales el sello de robustez y entereza que consigo trae el sexo masculino cuando esté bien caracterizado, y la docilidad y debilidad de carácter que domina en el sexo femenino.

We can see in his or her character and moral qualities signs of robustness and rectitude which are present in well-presented cases of masculinity, as well as the docility and weakness of character that predominate in the female sex.

In moral terms, as was the case of Reyes Carrasco, there is some doubt. In the case of N.N. instead of male qualities there is 'docilidad en sus acciones y debilidad de carácter peculiar a la mujer' (docility in her actions and weakness of character peculiar to women). He is 'extremely meticulous' in everything he does, he blushes frequently, is delicate in conversation and has female desires (understood not necessarily as sexual desires but general female practices). The very core of the individual is ambiguous: 'si su constitución física no se aviene mal con la reja y el arado, su carácter moral se encontraría mucho mejor con la aguja y la rueca' (if his physical constitution conforms well with the hoe and the plough, his moral character would be more comfortable with the needle and distaff).

The genitalia, however, would decide the case. As the genitalia were understood to be neutral, showing no evidence of being functioning parts, despite the ability for erection, and the individual showed no 'apetitos venéreos' (venereal appetite) or 'deseos eróticos' (erotic desires) of any kind, the case was confirmed as that of a 'neutral hermaphrodite'. In this diagnosis we see how the systems of analysis became extremely complex and individualized towards the late nineteenth century. Evidently, there was emphasis placed on the genitalia as the unequivocal sign of true sex, but a wide range of other characteristics were alluded to, including secondary and tertiary qualities. Furthermore, the *moral* qualities, understood as the 'proper' gendered lived experience or habitus of the individual were mixed in to an account that steadily grew less secure as more details were explored. Medical doctors were still bound by the need to harmonize the relationship between the body, 'sex', gendered habits and 'moral' criteria. That these criteria reflected a variety of classificatory systems (and indeed contradicted some elements of others, for example, the warnings that Marc made on the predilections of the 'patient') is clear. In addition, we can say that the divisions established by Klebs and Pozzi, for example, have not yet arrived in the medical consciousness as depicted here.[92]

Towards the age of the gonads?

We have argued that the period from the early 1870s up to the middle of the first decade of the twentieth century constitutes one of the crisis points in the science of hermaphroditism,[93] if not so much in the textbooks, certainly in the discussion of individual cases. In fact, the proliferation of individual cases grounded this ongoing re-evaluation and reports on them sought to move beyond some of the generalizations in textbooks. Instead of a clear, unified trend, the replacement of the use of 'hermaphroditism' by pseudo-hermaphroditism, apparent hermaphroditism or a different diagnosis such as hypospadias was a slow contested process that reached a crucial point around the late 1870s and 1880s. During this period different analyses coexisted. Furthermore, while interest picked up once more in sex identification from the early 1900s onwards, probably in response, at least in part, to new techniques such as laparotomy, there was a period from approximately 1890 to 1905 when medical journals were much quieter on the subject at least in terms of their attention to local Spanish cases. In the absence of Spanish examples, however, the period did witness a limited amount of reporting of foreign cases of hermaphroditism, pseudo-hermaphroditism and of hypospadias during the period 1890–1905.[94]

Such a development would entail a paradigm shift of the variety discussed by Kuhn – not in the sense that suddenly understandings were changed unrecognizably but that new frameworks were adopted that reflected new understandings already accepted as reality. Instead, following Kuhn, of theorizing advances in science as the motor behind change in themselves, a posteriori, these changes cannot explain the advances in science that have taken place. As Harding puts it: Kuhn claimed that the 'conceptual distinctions thought responsible for the great achievements in the history of science were in fact theorized only after the achievements had already been legitimated'. Kuhn also showed that these conceptual distinctions could not even in principle account for the historical processes they were intended to explain; historians 'had credited with the production of scientific revolutions the cognitive structures and inquiry processes that the revolutions only subsequently brought into existence'.[95]

Dreger has argued with respect to the arrival of Klebs's theory that it was only in the 1890s that his work became known in France and Britain, somewhat later, we might add, than Spain, where it was first reported in 1882.[96] The arrival of Klebs's theory that the gonads were the marker of the true sex of the individual, Dreger explains, did not in itself force a revolution in perception but reinforced a state of affairs already existing in those countries where Kleb's thought was present:

> Interestingly, Klebs's classification system apparently was not the catalyst for the shift in Britain and France toward the gonadal definition of true sex in the 1870s and 1880s, because neither Klebs' ideas nor his classification system seems to have been cited until the 1890s in any of the British or French literature. Instead it appears that the ultimate adoption of Klebs's classification system in France and Britain in the 1890s occurred simply because Klebs's system codified a by then already widely accepted notion, that is, that the gonads alone mark true sex.[97]

It is quite possible that notions that allowed for the revision of hermaphroditism, such as 'apparent' or 'pseudo-hermaphroditism', were present in Spain *avant la lettre*.[98] Such a state of affairs would codify 'a by then already widely accepted notion', to use Dreger's words on Klebs, that real hermaphroditism did not in fact exist but that, instead, a variety of sub-classifications along the lines of those suggested by James Young Simpson and Pozzi had received some acceptance. Simpson in 1839 broke hermaphroditism down into 'spurious' and 'true'. In the former case, the female variety would include 'excessive' development of the clitoris, adhesion of the penis to the scrotum and hypospadias. True hermaphroditism would be found in individuals who combined the genitalia and/or testes or ovaries of both sexes.[99]

An illustration of the complexities of this new conjunction, and its lack of linear progress amongst doctors involved in sex determination, is shown by the fact that in 1882 Klebs's classificatory system is mentioned, apparently for the first time in Spain, in a translation of a work by the Professor of Legal and Anatomical Medicine in Vienna, Eduard R. von Hofman.[100] Evidently, this apparent first citation of Klebs would not have entailed a thorough and instantaneous revision of current thought. The very same year, in a discussion that reveals that the relative importance of the gonads in sex determination was already a subject of controversy, a Dr Robert, reflecting on a case of hermaphroditism discovered in Barcelona, used the microscope to examine both spermatozoa and gonadal tissue.[101] In a review of the case the following was noted on Dr Robert's discussion about the sex to which the individual examined should belong:

> si solamente por la existencia del órgano que caracteriza el sexo, ovarios o testículos, había de deducirse a cual pertenecía el caso en cuestión, afirmaría que al masculino, pero sí, como él [Dr. Robert] opinaba y sostenía, el sexo debe caracterizarse por algo más que por la presencia de aquellos órganos, en el caso en cuestión se trataba de un hermafrodismo femenino.[102]

if the sex to which the case in question belonged were to be deduced solely from the existence of the organ that characterizes the sex, he would affirm the masculine. But if, as he [Dr Robert] believed and argued, sex should be classified by something more than the presence of these organs, the case in question would be classed as one of female hermaphrodism [i.e. essentially female but with certain male characteristics].

It is this 'algo más' that returns us to the classificatory systems of Marc and St-Hilaire. Even by the end of the century, it was possible to continue to publish medico-legal treatises in Spain which made no reference to Klebs's system or to the predominance of the gonads over other signs for the identification of sex.[103] Finally, it is worth noting that even Mata, in the sixth edition of his *Tratado* of 1903, maintained his text more or less unchanged. The same prominent international cases were studied; the N.N. couple study was reproduced as was that of R.V. é I. Barring some extra cases studies and data, there was little further elaboration of the subject of hermaphroditism in this classic textbook of legal medicine.[104]

The view from the body: pathological anatomy and physiology

We finish this chapter with a brief glance at another area of medical expertise: pathological anatomy and physiology. Medico-legal practice depended on an increased and more sophisticated knowledge of the body and its functions, developments that were associated with changes in medical practice, as we have noted, from the late eighteenth century onwards. The study of the body in the field of anatomy provided legal medicine with the necessary physical basis upon which to develop its forensic judgements. In turn, medico-legal doctors became slowly integrated into the structures of the state, being present in courts from the early nineteenth century onwards, a position largely unchallenged until the incursion of psychiatry at the start of the twentieth century.[105] In the field of pathological anatomy in Spain it is perhaps not unsurprising that the bodily characteristics of 'hermaphrodites' remained to the fore.

Schiebinger has noted that 'Science is not a cumulative enterprise; the history of science is as much about the loss of traditions as it is about the creation of new ones'.[106] The eclectic and innovative clinical text written by José de Letamendi, professor of general pathology and dean of the faculty of medicine at the University of Madrid, would certainly fall into the category of a lost discourse. Letamendi wrote his *Curso de clínica general o canon perpetuo de la práctica médica* in 1894 and the text was already somewhat of an anachronism in that, despite an increasing positivism in perceptions of life and nature, he defended a holistic and romantic conception akin to German *Naturphilosophie*.[107]

Letamendi divides the 'sexual impulse' into 'natural aphrodism' and 'para-aphrodism'. The latter category includes 'praeternatural or aberrant' sexuality. 'Natural aphrodism' follows the logics of attraction and repulsion, or sympathy and antipathy.[108] The 'para-aphrodisias', however, are expressions of praeternatural or aberrant sexuality and are explained as returns to an original hermaphroditic state, to a lack of sexual differentiation characteristic of the lower species and the first stages of individual development.[109]

These innate 'hermaphroditic remains' would be broken down into several different categorizations.[110] In accordance with the degree of atavism present in these 'erotic aberrations' there would be: (1) innate conditions as a result of anatomical structure; (2) innate conditions due to neurotic inheritance; (3) those due to spontaneous vice; (4) those resulting from bad example; (5) those due to pure necessity; and, (6) those produced by auto-suggestion and caprice.

Amongst 'homoerastic' subjects there are those who 'are naturally so' and those who are 'so through passion and those occasional'. Those who can effectively choose respond to 'secondary atavism or incomplete hermaphroditism (crossed para-aphrodism)'.[111] In this way, Letamendi coincided with the modern notion of 'inversion' as 'psychic hermaphroditism', divorced from any anatomical abnormality while maintaining the category of hermaphroditism heuristically to denote sexual deviation. Same-sex desires, in this sense, were *produced* by deviant anatomy,[112] which in Letamendi's formulation was distinctly hermaphroditic.

This reluctance to acknowledge the existence of hermaphrodites was reflected in other texts related to anatomical studies published in Spain in the early twentieth century. For example, in the manual on surgical techniques by Pierre Sebileau and R. Pichevin, *Tratado de cirugía clínica y operatoria* (1902), Pichevin dedicates just one page to hermaphroditism itself and, like all the authors reviewed above, remarks that 'La mayoría de los casos descritos con el diagnóstico de *hermafrodismo verdadero* no son otra cosa que *seudo-hermafrodismos*' (The majority of cases described with the diagnosis of 'true hermaphrodite' are nothing other than 'pseudo-hermaphrodites').[113] Two categories follow on from this description – gynandrynes and androgynes. The sources that the French author mentions represent, however, a break with other texts that we have discussed to this point. Pichevin refers to an 1898 edition of Pozzi's *Traité de gynécologie* (1898) and Neugebauer's *Verh. der Deutschen Gesellschaft für Gyn.* (1895) and to an unidentified source by Nagel. Relying on these sources the author is able to remark that the presence or otherwise of the 'sex glands' is a telling sign for real hermaphroditism.[114] This statement is followed by reference to Nagel and Pozzi as key resources for their identification.[115]

There is a difference, however, with respect to Pozzi's understanding of hermaphroditism and the role of identifying signs such as hypospadias. In a further disengagement of hypospadias from hermaphroditism, Pichevin treats hypospadias under general malformations of the vulva and vagina. In Pozzi's system, we will recall, hypospadias is classified as evidence of 'pseudo-hermaphroditism', as was the case in Simpson's classification.[116] Despite this admission, Pozzi is still working within the paradigm of hermaphroditism as the title of the section illustrates. Pichevin has removed hypospadias from the field of (pseudo-)hermaphroditism. He remarks with respect to male and female hypospadias that the latter is due to 'una suspensión del desarrollo de los órganos genitales, que recuerda la edad en que se encontraba el embrión en estado indeterminado' (a suspension in the development of the genital organs, which stems from the period when the embryo was in an indeterminate state), in accordance with a model of arrested development from a basis of poor sexual differentiation. The author noted, nevertheless, that this lack of development may be the cause of confusion with hermaphroditism itself: 'á esto se debe que se crea tan á menudo hermafroditas á los hipospádicos de uno y otro sexo' (it is because of this that so often hypospadiacs of one or other sex are believed to be hermaphrodites).[117] This text on surgical methods represents a dual shift. On the one hand, hypospadias is disconnected from hermaphroditism; on the other, explicit importance is given to the sex glands in the determination of sex. In this sense, Sebileau and Pichevin's *Tratado* is a taste of what was to come in the early twentieth century.

A half-way house position was adopted by other texts on medical hygiene and pathology in the early twentieth century. The pathologist Roberto Novoa Santos, in his *Manual de patología general*, acknowledged that there was a certain amount of natural variation of human and animal organisms in accordance with climate, work and existing pathogens. Novoa allowed for pathologies of the endocrinological system as a result of alterations in the 'sexual glands'. After mentioning virilism, the effects of castration and early puberty, Novoa briefly described three groups of hermaphroditism. The first included cases of lateral hermaphroditism, described by zoologists. The second referred to cases of true pure hermaphroditism, where glands of both sexes were present. These cases were unknown in man. The third set, which corresponded to the majority in humans, were cases of male and female pseudo-hermaphroditism. Within this category, there were three classes: 'secondary pseudo-hermaphrodites', where the individual contained the glands and sexual ducts of one sex and the secondary sexual characters of the other sex. There were also 'internal pseudo-hermaphrodites' where the internal sexual ducts present the characteristics of the sex contrary to that possessed by the subject in terms of his or her sexual glands. In a third class, the 'external pseudo-hermaphrodite', external sexual ducts contrasted with the primary sexual character of the individual.[118] As can be seen, Novoa relied on a classification system stemming not from Marc, who detailed apparent female hermaphroditism, apparent male hermaphroditism and neutral hermaphroditism, but not entirely from Klebs either. Klebs defined 'true' hermaphroditism as bilateral, unilateral and lateral. But Klebs did write about masculine pseudo-hermaphroditism with the presence of testicles and evident development of the female parts, and feminine pseudo-hermaphroditism with the presence of ovaries with some predominance of the masculine genital parts.[119] Furthermore,

Novoa's understandings did not stem obviously from Pozzi's system either, even though there are similarities.[120]

The steady elimination of real hermaphroditism and the elaboration of ever more complex systems of classification by the fields that we have discussed, such as legal medicine, general pathology, clinical treatises, surgical manuals, gynaecology and sexology, took place in lengthy textbooks and treatises. The picture in the clinic and the courts, however, seems to be somewhat different as cases of hermaphroditism reviewed were relatively frequent. Was it a case of these fields and disciplines professing in a more public and formal sphere – treatises and textbooks – that hermaphroditism did not really exist only to defend their right in the closeted sphere of the courts to pronounce on possible cases of hermaphroditism? The next chapter will focus on further changing paradigms whereby cases of ambiguous sex were recounted by medical doctors. Understandings of hermaphrodite science were to be revolutionized by emerging hormonal accounts, articulated in Spain in particular by Gregorio Marañón.

Notes

1 Orfila (1847: vol. I, 188). This version was translated from the fourth French edition and was updated in accordance with the Spanish legislation of the time by Dr Enrique Ataide from the Madrid Faculty of Medicine. Orfila was made Professor of Legal Medicine at Paris in 1823, according to Sánchez Ron (1999, 100, n.3). Orfila used the term 'hermaphrodism', which was more common in French than in British circles. Spain followed the French tradition up to the 1920s when hermaphrodism and hermaphroditism were used interchangeably. See Dreger (1998: 246, n. 9).
2 Cleminson and Vázquez García (2007).
3 Dreger (1997: 46–66).
4 Oudshoorn (1990, 1994).
5 Dreger (1998: 86, 93). Laparotomy was first used in France for diagnosing sex in the early 1890s but did not become a commonly used technique in France and Britain until the second decade of the twentieth century in those countries.
6 Ibid., pp. 139–66.
7 In 1878 the prominent figure in the field of legal medicine, Teodoro Yáñez, insisted on individual diagnosis as classificatory systems began to break down. See Yáñez (1878). Despite this emphasis, he still relied on Marc's classifications and referred to hermaphroditism as a 'detención' and 'retraso' of development (ibid., p. 294).

8   Glick (1976).
9   On this process see Foucault (1989: 124–48) and Ackerknecht (1967: 47–58).
10 De Letamendi (1894). Letamendi, according to Sánchez Ron (1999: 59), provided the first detailed discussion of Darwin in Spain in 1867.
11 See Foucault (1990a: 140–4); Dean (1994, 1999).
12 We have already referred to this in Ch. 1. Numerous examples of these concerns are reflected in medical journals throughout the period. See the medical criteria for exemption from military service reproduced in Anon. (1842: 210–13).
13 Aresti (2001: 137–43) focuses on the relations between masculinity, values and productivity in the first third of the twentieth century.
14 Schiebinger (1993: 10).
15 Ibid., original emphasis.
16 Labanyi (2000); Kirkpatrick (1989: 37–61); Aldaraca (1991); Jagoe (1994).
17 Aldaraca (1991: 66). The different role of women in the construction of the home and the education they should receive is analysed extensively in Di Febo (1976).
18 Jagoe (1994: 15–16).
19 Ibid., p. 17.
20 Ibid., pp. 36–7; Mangini (2001: 25–26).
21 Jagoe on marriage manuals (pp. 20–1); on psychiatry (p. 37). As we shall see, medical doctors investigating cases of hermaphroditism often relied on the notion of the female as inherently weaker, physically and morally, as well as being predisposed towards certain tasks.
22 Ibid., p. 37.
23 By examining the contents of the *Revista de Medicina y Cirugía Prácticas* from 1879 to 1910 this kind of evolution can be traced. For example, in this review, there were two cases of pseudo-hermaphroditism reported (one in Berlin and one in Valencia) between 1879 and 1890 and six cases between 1907 and 1910 (all in Berlin, Paris or Vienna). Cases of hypospadias or epispadias (where the urethra opens on top of the penis) follow the same pattern: two were reported between 1879 and 1890 (both foreign) while ten were reported between 1907 and 1910 (three of which were Spanish). This review has been a mine of information, whereas, for example, the *Anuario de Medicina y Cirugía Prácticas* (Madrid) from 1861 to 1881 and the *Anuario de Medicina y Cirujía* (Madrid) from 1885–1896, despite being very internationally oriented, produced no hits for hermaphroditism.
24 On the development of the medico-legal field in Spain see Martínez Pérez (1988).
25 Foderé (1801–3 (1798)). Before the Frenchman's text there was the slightly earlier Fernández del Valle (1796–7).
26 Cf. Peset and Peset (1975: 80–1).
27 Orfila (1847: 3), emphasis in original.
28 Orfila (1847: 5). In fact, according to Foderé (1813, vol. 1, xliii), there were three divisions of legal medicine: mixed, criminal and

sanitary. Although Foderé rejected real cases of 'androgyny', he did acknowledge that monstrosities had given rise to errors. By means of a visual and physical examination, doubt could be expelled and marriages could be annulled on the basis of infertility ('impuissance') (ibid., 355–66). In vol. 2 (178–9), doubtful sex in infants is discussed but this usually arises because of the lack of precision and knowledge of 'sages-femmes illitérées'. It was not the case that 'femmes instruites' did not exist; the people wanted ignorant women instead (pp. 179–80). Dreger (1997: 52) notes that Garnier in 1885 blamed 'ignorant matrons', 'prudish midwives' and 'myopic doctors' for mistaking the true sex; Dreger (1998: 215, n. 20) notes a similar judgement in 1894, whereby a hypospadic male's condition 'fooled' the matrons of a Brianconnais mountain village. A similar preoccupation was recorded in Seville in 1908. In the review of a report by Dr Antonio Morales Pérez originally in the *Revista Médica de Sevilla*, the doctor stated that it was usually the midwife that examined the new born and the family accepted her designation of male or female. The problem was that 'muchas veces no solamente ignoran estos problemas de hipospadias, si que también en la parte tocológica tienen una instrucción muy rudimentaria y empírica'. See Sarabia (1908: 447).

29 Orfila (1847: 163–99).
30 Ibid., p. 165.
31 Ibid., p. 166. In a later section, under 'De la impotencia' (pp. 170–83), it was pointed out that the lack of a penis or testicles, the existence of hypospadias or organic vices of the genitalia, could be causes of impotence. *Impotence* may not be a just cause for the annulment of marriage, but *infertility* continued to be so.
32 Ibid., pp. 169–70.
33 Ibid., p. 188.
34 Marc (1817: 114).
35 See Comenge (n.d.) and Sebileau and Pichevin (1902), discussed below.
36 Orfila (1847: 192–3).
37 Ibid., p. 193.
38 Mata was responsible for the 1843 Education Reform Act, which set up Chairs of Hygiene in Spanish universities. See Labanyi (2000: 69–70).
39 Mata (1844, vol. 1, pp. 15–21).
40 Ibid., pp. 15–16.
41 The same system of classification, with the designations masculine, feminine and neuter hermaphrodism, were recorded by Cayetano del Toro y Quartiellers (1876, vol. 1, pp. 430–2). Hermaphrodism for Del Toro was a kind of 'monstrosity', following St-Hilaire.
42 Mata (1874: vol. 1, pp. 346–7) records these same characteristics, including the same lack of inclination towards the female sex as a characteristic of the masculine hermaphrodite.
43 Also referred to by Yáñez (1878: 291).
44 Mata (1857, vol. 1, p. 265).
45 Ibid., p. 267.
46 This case is reproduced in Mata (1874, vol. 1, pp. 356–9).

Between Diagnoses 117

47 Towards the latter part of the nineteenth century the notion of identity would alter to become more personalized and more a designation of the interior, intimate realities of the person or subject.
48 Mata (1857, vol. 1, p. 274).
49 Mata (1874, vol. 1, pp. 291–761).
50 Ibid., pp. 339–63.
51 Ibid., p. 340.
52 Ibid., pp. 345–6. Mata was certainly ahead of his time in discussing Alexina. Amancio Peratoner, otherwise up to date in his sexological knowledge, does not mention Alexina in his chapter on hermaphroditism in *Higiene y fisiología del amor en los dos sexos* (1880: 233–252).
53 Mata (1874, vol. 1, pp. 341–4).
54 Ibid., p. 342.
55 Ibid., pp. 360–1; Alba y López (1860: 265, 1861).
56 Mata (1874: 360); Alba y López (1861: 455).
57 Mata (1874: 360).
58 Alba y López (1860: 265).
59 Alba y López wrote of 'el hombre ávido de ciencia' (the avid man of science), 'ese acto de filantropía' (that act of philanthropy), 'la ciencia más noble' (the most noble science) and 'el humano saber' (human knowledge) (1860). The irony, of course, is that had the patient not 'consented' we would not now, 150 years later, dispose of such materials from which to draw our analysis. Further quotations in this section derive from the three texts already mentioned.
60 Hypospadias of the third degree was most likely to be classed somewhere along the axis of hermaphroditism. By the early twentieth century, this was all that remained of the association between the two, and it was rapidly eroded. This development is discussed more fully in Ch. 4. An example can be seen in Morales Pérez (1906). This case was reproduced in *RMCP*, 79 (1908), 446–8.
61 On this point see Dreger (1997: 56, 63, n. 8).
62 Chase (1998: 189).
63 Perez and Cherizola (1864), discussed in detail in Vázquez García and Moreno Mengíbar (1995b; 1997a: 215–19).
64 Dreger (1997: 57) notes that many medical figures believed that same-sex desires were literally produced by deviant anatomy. Similar understandings prevail in the account by Pérez and Cherizola.
65 Perez and Cherizola (1864: 74).
66 Legrand du Saulle et al. (1898: vol. 2). This book was based on Legrand du Saulle's previous volume of the same title published in 1886 and annotated in accordance with Spanish and Latin American legislation. It was translated by Drs Teodoro Yáñez and Núñez Granés. Note José Núñez Granés translated Sebileau and Pichevin (1902). Mata (1874) notes that he follows Du Saulle, amongst others, in order to help classify hermaphrodites. Legrand du Saulle et al. (1898: 499–572) discuss in Ch. 7 the topic under 'Matrimonio'. The volume discusses marriage legislation in Spain, the changes that took place in 1870, allowing civil marriage, and the annulment of that law in 1875

which reaffirmed church marriage alone. Subsequently, civil marriage was once more permitted under the law.
67 Legrand du Saulle et al. (1898: 527–33). It will be recalled that Mata distinguished between errors of identity (where one of the couple was not who they were purported to be) and errors of person or sex (in the case of vices of conformation or hermaphroditism).
68 Ibid., p. 531.
69 Marc (1817: 116): 'Enfin on ne devra tirer parti qu'avec une certaine réserve des déclarations de l'hermaphrodite ou des personnes qui ont des liaisons directes avec lui' (one should only draw upon the declarations of the hermaphrodite or persons who have direct contact with him or her with a certain degree of reservation).
70 Legrand du Saulle et al. (1898: 531).
71 Ibid., p. 532.
72 See Dreger (1998: 142).
73 Legrand du Saulle et al. (1898: 532).
74 Ibid., p. 533.
75 J. M. M. (1888).
76 Briand et al. (1872–73: vol. 2, p. 186). See also Casper (1886–7: vol. 5).
77 Legrand du Saulle et al. (1898: 531–2).
78 Peset Cervera (1879). It had been argued in Anon. (1865) that 'Hasta llegar á la pubertad no hay verdadero sexo; hasta entonces no hay mas que niños; los organos genitales no son órganos de poder, sino de espera' (Until arriving at puberty there is no true sex; until then there are just infants. The sexual organs are not potent but in waiting).
79 Dreger (1997: 56, 63, n. 8).
80 Anon. (1888).
81 Pozzi (n.d.: 182): 'the field of hermaphrodism is becoming narrower by the day'.
82 There is evidence of attention being given to these techniques from abroad. News from the French Association for the Advancement of the Sciences on this subject was detailed in Anon. (1879), and previously, from France in Anon. (1861).
83 Despite this, influential figures still included hypospadias in their classifications of pseudo-hermaphroditism. See Pozzi (1893: 456–63), although there was the admission that 'the great majority of the pseudo-hermaphrodites which have been described and pictured are men with hypospadias' (p. 452). The Spanish version (Pozzi, n.d.: 585–92) followed exactly the same criteria.
84 Galcerán (1877). The case was reported by Sr Esquerdo to the Sociedad Médica 'El Laboratorio'.
85 Ibid., p. 178. A similar case was reported in Paris at the Society of Surgeons in 1881 where the diagnosis varied between hermaphrodism and hypospadias and was commented upon in 'Hermafrodismo ó hypospadias' (ibid., pp. 181–2). An individual was thought female, married, but was seen to have a 'twelve-year-old boy's penis', with testes but no sperm. According to St-Hilaire, he would be an imper-

fect bisexual hermaphrodite, that is, from the Second Class, Order Three, as in Dreger (1998: 142), the doctor describing the case noted. If recent embryological research was taken into account, however, 'se llegaría á esta conclusion de que en este caso se trata en efecto de un hombre con hypospadias' (one would arrive at the conclusion in this case that what we have is indeed a man with hypospadias). Pozzi himself was present at the session and confirmed this diagnosis, noting 'El campo del hermafrodismo se va estrechando cada vez más; no se ha presentado hace veinte años un solo hecho de hermafrodismo bien demostrado: creo que respecto á este punto debe ser uno muy escéptico' (The field of hermaphrodism is becoming more and more narrow; for twenty years no single example of clearly demonstrated hermaphrodism has come to light. I believe one should be sceptical on this question) (Galcerán, 1877: 182). M. Tillaux spoke up in defence of 'real hermaphrodites ... who have organs of the two sexes, testicles and ovaries, at the same time' (p. 182). The piece ended with the comment that hermaphrodites were only those individuals who possessed the glands of both sexes.

86 Mariani (1880). The case was reported at the Academia Médico-Quirúrgica Española.
87 Anon. (1880).
88 Dreger (1998: 142).
89 Ibid.
90 See James Young Simpson's 'Hermaphroditism, or hermaphrodism', in Robert Bentley Todd (ed.), *The Cyclopaedia of Anatomy and Physiology*, vol. 2 (London, 1839), pp. 684–738, as discussed in Dreger (1998: 142). Young's article was reprinted in 1872.
91 As discussed in Pozzi (1893: vol. 3, p. 463).
92 That various diagnoses and cognitive systems coexisted simultaneously is further illustrated in Mariani (1886: 41). This case of a prisoner was originally published in *La Crónica Médica* by Dr Ricardo Mariana y Albiol, medical doctor in the Serranos (Valencia) prison. The ambiguous subject in this case should be placed in a room separated from one and other sex as he possessed 'los atributos del masculino con algunos del contrario' (the attributes of the male with some of the opposite [sex]). Even though the harmony between sex, body and gender was thus restored, the in-between figure remained a dangerous individual in the ordered world of late nineteenth-century society.
93 Vicente Cacho Viu argues that the 1898 crisis caused by the loss of the Spanish colonies was reflected in a crisis of positivism in Europe at the end of the century in that reason and science alone were not deemed sufficient to explain the world. See Cacho Viu (1997: 58–62).
94 For example, see Anon. (1891), 'Un caso de hermafrodismo', from the Paris Medical Academy; Anon. (1890), 'Pseudo-hermafrodismo masculino', from the Berlin Medical Society, relates the case of a 24-year-old who was brought up as a girl up to the age of 12 when she changed dress and social role and expressed desire towards women.

While some aspects were womanly, he possessed an imperforate penis and had a marked double cryptorchidism.
95 See Kuhn (1970); Harding (1986: 199).
96 See Hofman (1882: 49). Both the English- and Spanish-language edns of Pozzi's *Treatise* mentioned Klebs's divisions.
97 Dreger (1998: 146).
98 Indeed, an early elaboration of this was that of Peratoner who declared that there were two types of hermaphroditism: male and female pseudo-hermaphroditism. See Peratoner (1880: 234–5, 250–1). In fact, the notion of pseudo-hermaphroditism as such in Spain only caught on in the 1920s, as we shall see in Ch. 4.
99 Dreger (1998: 143–4), citing Simpson's article of 1839.
100 Hofman (1882: 49). This text was translated by Dr Carreras Sanchís and dedicated to Dr Mata who had died five years previously. It is interesting to note that hermaphroditism was not explicitly discussed, nor was Klebs, in Hofman's later *Tratado de medicina legal* (1891). We have only been able to consult vol. 2 of this publication. However, a case from 1886 petitioning the annulment of marriage is discussed under ecclesiastical law and reproduced in this volume with a commentary by Dr Adriano Alonso Martínez (Hofman, 1891: 506–9). Instead of classifying the subject in question as a hermaphrodite Dr Alonso argues for the annulment of the marriage on the basis of hypospadias and impotence. This constitutes another example of the disappearance of the hermaphrodite and his or her explanation by other medical means.
101 The use of the microscope for various medico-legal procedures had been discussed throughout the nineteenth century. It was advocated for the testing of sperm in cases of suspected hermaphroditism and/or impotency in a brief assessment of Orfila's *Del esperma considerado bajo el aspecto médico-legal* in Anon. (1841); in Anon. (1863) on the subject of Casper and Tardieu's work (Anon., 1863), and in Mariani (1886).
102 Robert (1882). Dr Bartolomé Robert y Yarzábal was a member of the Facultad de Medicina at Barcelona.
103 The influential *Tratado de antropología médica y jurídica* by Ignacio Valentí Vivó (1889: 397), noted that 'lo sobrenatural, maravilloso o absurdo es totalmente inadmisible en este capítulo de la Embriogenia humana' (the supernatural, marvellous or absurd is completely inadmissible in this chapter of human Embryology) but did not mention Klebs.
104 Alexina's account was published in the 6th edn of Pedro Mata's *Tratado teórico-práctico de medicina legal y toxicología* (1903: 317–18).
105 On the integration of psychiatry into the courts in Spain see Campos Marín (1999a, 1999b).
106 Schiebinger (1989: 2).
107 We discussed Letamendi more extensively in Cleminson and Vázquez García (2007: 50–3).
108 De Letamendi (1894, vol. 2: 110).

109 Ibid., p. 119.
110 Ibid., p. 120.
111 Ibid., p. 126.
112 The proximity of homosexuality to hermaphroditism and 'psychic hermaphroditism' has been discussed in Ch. 1. A Spanish example is Bayo (1902: 109) where it is remarked that the popular term for pederasty and Sapphism was 'hermaphrodism'.
113 Sebileau and Pichevin (1902: 672).
114 Ibid.; emphasis in original. Pichevin wrote this section of the co-authored volume. Samuel Pozzi (1893: 452–67) refers to the 'genital gland' in the context of a discussion on ovaries and testicles in a section on 'Hermaphrodism'.
115 Sebileau and Pichevin (1902: 672). Pichevin also remarks that pseudo-hermaphrodites often suffer from psychological problems, 'desórdenes en la esfera psíquica' (disorders in the psychic realm, p. 672).
116 This trend was continued by Recasens (1918: 384–6), where hypospadias and epispadias are not considered under hermaphroditism; indeed, there is a separate section on the latter (pp. 386–8). Two reports on epispadias in *El Siglo Médico* from the late nineteenth and early twentieth centuries do not discuss hermaphroditism. See Iglesias y Díaz (1892). Here Dr Alejandro San Martín reported on a case of third degree epispadias in a 12-year-old boy from Córdoba. See also under the foreign medical periodicals section by Navarro Cánovas (1904).
117 Sebileau and Pichevin (1902: 669).
118 Novoa Santos (1916: 414–17).
119 All correct as in Dreger (1998: 145).
120 Pozzi (1893, vol. 3: 453).

Chapter 4
Gonads, Hormones and Marañón's Theory of Intersexuality, 1905–1930

In 1959, at the chronological limits of this study, Gregorio Marañón, the 'Spanish Darwin'[1] who espoused 'doctrinas en pro de la feminidad y varonía correctas' (doctrines in favour of correct femininity and masculinity),[2] presented a case of female homosexuality with male chromosomal sex in the bulletin of the Institute of Pathological Medicine.[3] This case was presented in a similar way to those of 'hermaphrodites' earlier in the century. 'Magdalena S.' was 35 years old, was single, had been born in Seville and was currently employed as a shop-worker in Madrid. Three cousins, like her, were 'amenorreicas' (suffered from amenorrhoea).[4] There was a further female relative who was considered a girl until the age of 10 and who then changed clothing and began work as a man, described by Marañón as a case of transvestism. Magdalena, like other women in her family, produced scant menstrual fluid and possessed 'Órganos genitales normales, sin aumento de clítoris' (normal genitalia, with no augmentation in clitoris size).[5]

She had attempted sexual relations with both men and women, especially the latter. An extensive and complex process of medical diagnosis, with Dr López Aydillo providing a neuro-psychiatric report, Dr Pérez Cuadra a general medical history and Dr Puchol a hormonal account, helped Marañón confirm that Magdalena was chromosomally male. This study also referred to a previous case discussed by Marañón along similar lines. This previous case of a male homosexual with female chromosomal sex was an 'hallazgo rarísimo' (extremely rare find). Other authors had found that transvestite homosexual men all possessed male chromosomes, which given the fact that Marañón's case was a male homosexual transvestite with female chromosomes made the example even more interesting.

In spite of the proximity of the diagnostic process to earlier accounts of hermaphroditism or pseudo-hermaphroditism,

Magdalena's case throws up important differences. Clearly, the techniques used to determine her homosexuality, transvestism and male chromosomes are different. In addition, the very category into which she was placed had little in common with those of thirty years before. There was even some doubt as to the subject's sexual preference given the question mark over her sex. Marañón conceded that it could indeed be argued that there was some doubt as to whether she 'really' was a homosexual after all given that, although she had undertaken 'homosexual' acts, 'En realidad es genéticamente un hombre con apariencia de mujer, si bien un tanto viriloide, con caracteres sexuales primarios muy de mujer y caracteres sexuales secundarios equívocos' (In reality she is genetically a man with a woman's appearance, although somewhat viriloid, with primary sexual characters very much of a female type and secondary sexual characters of an equivocal nature).[6]

Her lack of menstruation, 'vigorous muscular system', her height, 'robustness', 'scarce maternal element' in her mammary development, male skeletal arrangement and attraction to the female sex led Marañón to classify Magdalena as a typical virile female homosexual whose sexuality was closer to that of children and adolescents. This was despite the existence of the male chromosomes. Marañón affirmed 'que el sexo genético puede servir de dato clasificador' (that genetic sex can serve as classificatory data) for cases of intersexuality, hermaphroditism, pseudo-hermaphroditism, syndromes such as Turner's and Klinefelter's, as well as for 'las formas más interesantes y frecuentes de la intersexualidad, que son los estados homosexuales' (the most interesting and frequent forms of intersexuality, which are the homosexual states). But the 'genetic sex' of the person is one factor among many others, not least endocrinological and psychological dimensions.

What we might term 'social sex' was a major factor to be taken into account. Marañón goes on to remark that, from the social point of view, his patient 'acts like a woman'. Furthermore, despite some episodic heterosexual encounters, she displayed a preference for women and she was a woman in social terms. These social aspects and her sexual orientation 'autorizan a clasificarla como homosexual' (allow her to be classified as a homosexual).[7]

In contrast to some, but by no means all, cases of suspected hermaphroditism up to the end of the 1930s, the sexual preferences of the individual were played down and other criteria, as we have seen, were used to determine sex within a strict heterosexual model. Here, however, what we see is the emergence of new categories such as transvestism, chromosomal sex and even homosexuality entwined, where social sex and social sexuality are understood to be key – or even decisive – elements. This was a period when chromosomal sex, sexuality and gendered behaviour had become detached from dualistic sex-specific traits.

But Marañón's thought had not always displayed such diagnostic traits. Before the rise of a chromosomal explanation for real sex, Marañón had articulated a theory of intersexuality in the late 1920s that included a huge range of characteristics from virilized women through to male homosexuality and actual hermaphroditism. During this period when sexual identity in other countries was 'questioned, resolved and questioned again',[8] Marañón had argued, against the international grain in hermaphrodite science, that cases of 'pseudo-hermaphroditism' should be really labelled 'true hermaphroditism' as this best described what they really were.[9] In Britain and France the notion of pseudo-hermaphroditism was slowly eliminated. If hermaphroditism did not really exist in humans, how was it possible to have a 'pseudo' version of something that was non-existent?[10] In Spain, however, the category hermaphroditism experienced revitalization from the late 1920s.

This battle over the scope and definition of hermaphroditism in Spain in the early and mid-twentieth century was a product not only of Marañón's theory of intersexuality but also of the broader process of definition of the sexes – some doctors and social commentators wished to reassert difference between the sexes in the light of what was commonly understood as gender muddling by feminists, New Women and increasingly visible homosexuals at the time. This chapter takes our story from the prevalence of the gonadal model, discussed in Chapter 3, as an identifier of true sex, analyses the emergence of hormonal indicators of true sex and considers the expansion of the scope of hermaphroditism as espoused by Marañón from the late 1920s. In the early 1930s, a further crisis of thought enveloped hermaphrodite science within the context of new endocrinological understandings. Doubt entered the world of hormonal analysis when those hormones that were thought to be sex-specific, clearly showing real sex, were found to be present in both sexes. This situation would characterize studies up to the late 1950s when we see the emergence of two competing accounts: social sex and chromosomal definition. Both would participate in further reducing the discursive field of hermaphroditism, allowing for the emergence of a new category: transsexualism.

In the period covered in this chapter, it is possible to trace a number of tendencies for Spain in the science of hermaphroditism during the period 1905–30. First, we detect a continuation of the situation outlined in the previous chapter with respect to an ongoing silence on cases of hermaphroditism and on classificatory systems. This situation lasted up to the early 1910s. During this period, cases of ambiguous sex which might previously have been classified as different forms of hermaphroditism were presented as cases of hypospadias of differing degrees. While hypospadias of the genitalia was admitted by nineteenth-century physicians such as Marc and later by Pozzi as forming part of the category of hermaphroditism or for rendering classification more complex and equivocal, in Spain, discourse on hypospadias supplants almost completely that on hermaphroditism during this period. Such a tendency operated in tune with what was taking place on an international front and numerous examples of hypospadias discussed in foreign medical journals were reported upon in the Spanish medical press at this time. Secondly, we see a comparatively later uptake of the classification of pseudo-hermaphroditism in case studies in Spain *vis-à-vis* foreign reception (which was also reported on). Spanish doctors only began to classify cases as 'pseudo-hermaphroditism' from the mid-1910s onwards, when the gonads came to be accepted as the marker of true sex. In fact, it will be argued here that if there was an 'Age of Gonads' in Spain it was apparently consolidated later than in France and Britain or was at least more equivocal. Rather than a definitive or hegemonic stand on sex identification being held by the 1870s in accordance with the gonadal model, what we see is a plurality of positions taken by doctors, in which gender traits were understood to be valuable in the identification of true sex. This could lead us to think that the Spanish medical sciences were behind those of some other European countries. Instead, it is argued here that sex was partly expressed by gendered traits more than the physical presence of gonads for a longer period in Spain than in other countries. Such a hypothesis would coincide with studies on male homosexuality in Spain where 'sexual inversion' as a category only expressed gender *and* sexual deviance from the 1920s onwards, not in the 1890s or before as in the United States and Britain. This quality acted as a retardant in the acceptance of the term 'homosexuality' in Spain during the period.[11]

Thirdly, the 1920s were characterized by multiple explanations of hermaphroditism. During the early part of the decade it was accepted that the gonads were the sign of true sex, and we see the decay of the term hermaphroditism and the increased usage of the notions of apparent or pseudo-hermaphroditism. The latter terms had been occasionally employed in the 1870s and 1880s in Spain (see Chapter 3). At the end of the 1920s, this short-lived consensus was shattered by Marañón's notion of intersexuality, causing yet another reassessment of hermaphrodites forming a fourth tendency or characteristic of this period.

Throughout the years 1905–30 the acceptance of the importance of the glands and the internal secretions in sexual questions varied, generally gaining importance from an early exposition of their effects in the first decade of the 1900s to a more dominant, but never hegemonic position, in the mid-1910s and 1920s. In addition to tracing the evolution of medical thought on hermaphroditism, this chapter interweaves, at least for the 1920s, a glance at social change as a factor, which drove or at least informed doctors' notions (and concerns) about sexual difference. This, it will be argued, was particularly the case with Marañón, whose thought on sexual differentiation clearly responded to anxiety over changing gender roles and recapitulated what might now be termed essentialism in respect of gendered and sex roles, with 'biological' traits thought to play a large part in the conditioning of correct gendered behaviour. In accordance with this understanding any gender deviance was either pathologized or seen as exceptional.

Hypospadias or hermaphroditism? Different diagnoses, different life stories

In 1817 Charles Chrétien Henri Marc (1771–1840) emphasized what he called the 'vices of conformation' in human beings. These made identification of sex a somewhat problematic endeavour.[12] Marc described 'Apparent hermaphrodism in the male sex' as a 'vice of conformation of the scrotum', which divided the scrotum into two large labia with displacement of the testicles, a smaller penis than normal, a urethra that exited from the base of the penis and in some cases the rectum communicated with the penis and the scrotum. This was one of the most common classes of hermaphroditism.[13] In the case of apparent hermaphroditism in the female case, there were two sorts of vices of conformation that 'can induce error on the nature of their sex'.[14] The first of these was a large (although imperforate) clitoris. The second was the existence of a glans and prepuce on the clitoris and the lack of internal and external labia. In addition, the urethra was in a place where the vagina should normally be. Marc, in the section on the medico-legal application of this knowledge, names these individuals as hypospadiac, epispadiac and anaspadiac.[15]

Samuel Pozzi, in his influential *Treatise on Gynaecology* published in 1892–3 in England (translated from the French edition of 1890 and published shortly afterwards in Spanish), wrote extensively on the subject of hypospadias in his section on hermaphroditism.[16] In fact, Pozzi argued, hypospadias was viewed as the condition which most led to supposed cases of 'double sex'. Such a phenomenon would appear as a result of the vice of conformation of the genitalia in two cases: the under- or suspended development in the embryological phase and the excessive development of the genitalia. The first set of cases was the most common and most cases of pseudo-hermaphroditism were in fact, according to Pozzi, cases of hypospadiac men. These cases would form part of a multiple category of 'pseudo-hermaphroditisms'. Pozzi's ideas on this area will be examined in a later section when we discuss the rise of the explanation of pseudo-hermaphroditism in the mid-1910s. For now, it is worth stating that the exposition of these cases was extensive in Pozzi's *Tratado* and they were illustrated in the same way as the English-language edition.[17]

While Pozzi was only mentioned in passing during the late nineteenth century in Spanish doctors' accounts of hermaphroditism, an increased interest in the condition of hypospadias is evident from the beginning of the twentieth century onwards. It took some time for actual case studies to be recorded in accordance with this criterion. Nevertheless, there was ample reporting on foreign doctors' discussions of cases of hypospadias from 1900 onwards. It is possible that this arrival of increased interest in hypospadias both nationally and internationally supplanted diagnosis of cases as 'hermaphroditism', after the assault in the mid- to late nineteenth century on the existence of real hermaphrodites of any variety. A first explanation for the silence around hermaphroditism during this period would therefore be of a medical nature where discourses tended towards the resignification of a condition, perhaps as part of a struggle to utilize the latest criteria and terms offered by foreign specialists. Other reasons, however, may well respond to other discursive effects of a crisis in positivism and its related systems of classification and categorization, already mentioned in Chapter 3, in the light of the 'crisis' of 1898.[18] Such an interest could also be related to the growth of surgical procedures in Spain as many of the cases recorded were deemed possible to correct by medical intervention. Finally, the very real concern over visible diseases such as tuberculosis and syphilis, for example, may well have displaced interest in questions related to this particular branch of sexuality-related issues. In contrast, the rapid process of modernization, changing roles for women and emerging feminism would all aid a restructuring of what was understood to be sexual identity from the 1920s on.

Reporting of foreign cases and interest in hypospadias was constant throughout the period 1900–20 in Spain. These were reported upon in the sections of journals dedicated to examining the foreign medical press. For example, there were reports on a 'very rare' case of epispadias recorded in Germany in 1904,[19] and a case of doubtful sex from Berlin that was explained by the existence of a 'scrotal epispadias' in 1910.[20] Other examples include an analysis of treatment for cryptorchidism in Berlin,[21] more than one report on treatment for hypospadias as elaborated in Paris and two cases of hypospadias recorded in Vienna.[22]

Despite the occasional doubt over the sex of the patient, hypospadias was treated as a medical condition to be resolved, generally, by surgery. Hermaphroditism remained unmentioned in all these cases. The only exception to this was an explicit discussion by Pozzi himself on surgical techniques and hypospadias, appearing in the *Revista de Medicina y Cirugía Prácticas* in 1907. This was one of the rare cases when Pozzi was actually cited or quoted in the Spanish medical press before the 1920s. Dr Codina Castellví reported that at the Paris Academy of Medicine Dr Pozzi read a report on a case discussed originally by a Dr Barnsby.[23] Although the reporting of the case may well be one of the first mentions of the term 'pseudo-hermaphrodism' in the twentieth century in Spain the piece focused primarily on surgical techniques and the need for close inspection of the genitalia for the purposes of correct sex identification. Pozzi considered the case of a 12-year-old boy who had always been considered a girl. While there was a 'pseudo-vagina', 'pseudo-large labia' and one atrophied testicle, the imperforate penis confirmed the male sex. An operation was performed. The language employed implied apparent, pseudo or even false sexual characteristics, which could be correctly identified by means of the specialist's eye and surgical hand and brought into line with the true sex of the individual.

This collection of reports drawing on a wide variety of foreign medical academies and journals was fairly extensive in its coverage and was very much up to date. Accounts of scientific sessions in foreign institutions usually appeared within the same year or shortly afterwards in the next year in the Spanish journals. These show that 'hermaphroditism' is displaced in foreign medical accounts. In similar cases discussed during the later years of the nineteenth century the individuals examined would probably have been pronounced to be some kind of hermaphrodite and classified accordingly. In the early twentieth century, however, even though some cases of 'doubtful sex' proved somewhat controversial, any questions of unusual genitalia were promptly viewed as cases of hypospadias or vices of conformation. Hermaphroditism continued to fade away from foreign medico-legal accounts.

The same general tendency was played out amongst Spanish medical doctors. Although there appeared to be a greater margin of variability in the presence or absence of hypospadias and the recourse to the category of hermaphroditism in the cases reported, this may have been the result of the medium in which these local cases were analysed. We must recall that the foreign discussions were filtered at least twice – first by the foreign press itself and secondly by the Spanish commentator.

An elucidation of the differences and possible confusions between hypospadias and hermaphroditism was elaborated in the *Siglo Médico* in 1906. A year before the account given by Pozzi on 'pseudo-hermaphroditism' appeared, Dr Rafael Nevado Requena discussed a case of doubtful sex from the small town of Lucar.[24] The case discussed by Dr Nevado was that of a newly born baby who apparently possessed the genitalia of both sexes. The family and local priest had declared the baby to be a 'hermaphrodite'. Nevado reported on a healthy baby with 'ojos vivos y expresivos' (lively and expressive eyes).[25] What did appear to be amiss, however, were the genital organs. An imperforate 'pene de dos centímetros en estado de flacidez' (penis of 2cm in a state of flaccidity),[26] was viewed and was reported to be adhered to the raphe and perineum. The urethra was discovered at the base of the penis. It quickly emerged that the child possessed two testicles 'the size of hazelnuts', by which the sex was identified: 'Una vez determinado el sexo, se le puso nombre de varón' (Once the sex was determined, he was given a boy's name).[27] Apart from showing the dominance of the gonads in identifying the 'correct' sex of the baby, we also see the expulsion of any notion of hermaphroditism early on in the account. After the initial two paragraphs that related the views of the non-specialist family and priest and the general condition of the baby, the medical report proceeded apace to conclude in one slightly longer paragraph and a short declaration the fact that what was before the doctor, family and priest was in fact a baby boy.

This determination of sex was followed by an extensive account of the nature of hypospadias in this case and in general. This particular case reflected a hypospadias of the third degree whereby the scrotum was 'completely sunken', representing a vulva and making the penis appear to be a clitoris, giving the overall impression of male hermaphroditism. This was only an appearance, however: in fact what the doctor beheld was 'un hipospadias congénito, parafimosis y ectopia testicular' (congenital hypospadias, para-phymosis and testicular ectopia).[28] Dr Nevado explained that, in accordance with embryological theory, there were three stages in the development of humans: a period of a lack of sexual differentiation, a period of hermaphroditism and, thirdly, a period of sexual differentiation.[29] Any form of hermaphroditism was explained as emerging between the fourth week and the third month of gestation. What had happened in the Lucar case was a classic case of 'suspension of development', resulting in hypospadias.

In the same way as many other diseases and conditions were passed on, according to a degenerationist model, apparently hypospadias could also derive from a congenital source. We learn that the baby's great grandfather was hypospadiac; an aunt had to change her name from Ramón to Ramona. Ramona worked in the fields, was shy and retiring, complained about not being allowed to wear trousers and was in fact a case of female hermaphroditism. Her clitoris was 'excessively developed' and approximated an adult penis.[30]

Dr Nevado's authority was questioned, however, by the family (possibly under the influence of the priest) and they opposed any form of treatment. If they did finally consent, Dr Nevado noted, however, the usual surgical methods would be attempted, following the surgical techniques perfected by Dr Pulido Martín 'que con tan brillante éxito ha estudiado la especialidad en las naciones extranjeras' (who has studied the specialism with such brilliant success in foreign nations).[31] Dr Nevado finished his account with the admission that in the three years he had been practising medicine in the locality he had seen no other cases of congenital abnormality of the genitalia and urinary apparatus.[32]

The case is interesting because it shows, on the one hand, how the gonads were becoming the principal means by which the true sex of the individual would be identified. On the other hand, Dr Nevado's account shows the liminal status of 'hermaphroditism' in medical discourse at the time. Even though he accepts the cases described to form part of the categories of male or female hermaphroditism, the whole piece tends to play down this diagnosis in favour of 'apparent' hermaphroditism (an aspect reflected in the title) and hypospadias throughout. While, as we shall see, the category 'pseudo-hermaphroditism' would increasingly (re)gain importance as the 1910s wore on, here what counts is the malformation of the genitalia in dubious cases of sexual identity.

A cluster of cases from 1908 confirm this general tendency towards the supplanting of the diagnosis of hermaphroditism by hypospadias. Our first example relates the case of 'S.A.', 25, single, and an electrician from Madrid. While his hereditary antecedents are unknown, he is described as an 'alcoholic' and 'degenerate'. A brother and his fiancée suffer from similar genital irregularities. His testicles were small, as was the penis, which was visible only as the glans covered by the prepuce, which in turn 'se hallaba sumamente engrosado y replegado hacia atrás, en disposición parecida á la capota replegada de un coche milord' (was extremely swollen and folded backwards, in the same way as the roof of a convertible car).[33] The reporting physician, Dr Angulo, insisted that there was no hypospadias in this case and there were several places along the side of the penis where urine was expelled. The process of rebuilding the penis is then detailed. This operation was successful and the man's penis was 'pequeño, pero libre y saliente, apto ya para la cópula' (small but free and emergent, now capable of copulation).[34] Dr Angulo's denial of hypospadias in this case raised some doubts in his audience at the Madrid Surgical Society. Dr Royo insisted that the case was nothing less than one of hypospadias – he had also seen cases with the kind of lesions described by his colleague. Royo advocated operating as soon as possible, around the age of 7 to 14 years, in order to prevent the future melancholy of any young man in the same situation. Another doctor, Dr Barragán, asserted that Angulo's was a case of hypospadias, although it was unusual and Dr Parache entered into an embryological explanation of the condition. Finally, Dr Goyanes referred to at least one case of second (perineal-scrotal) or third degree (perineal) hypospadias operated on successfully in Spain to date. There was a final comment from Dr Angulo who protested that his case could not be understood as an example of hypospadias.

There was no similar degree of doubt expressed in the cases referred to by Dr Antonio Morales Pérez in the same year.[35] Here, Morales accepted that hypospadias of the third degree could be confused with hermaphroditism but his title alluded to the falseness of this claim; here he related other cases of hypospadias '*appearing* to be hermaphroditism'. The details were as follows: the first case was that of a girl of 5 years from Olot. Morales was called in to perform an unspecified operation and discovered by looking at the 'genital organs' that the girl was in fact a boy. The course advised by the surgeon was that the father 'le cambiase el nombre de hembra por el de varón, y lo visitiese y educase en las condiciones varoniles para evitar ulteriores *consecuencias sociales*' (should change the girl's name for a boy's, should dress and bring him up in manly conditions in order to avoid subsequent *social consequences*).[36] Nothing more was heard of the boy. From such a scant account it is difficult to extract any major conclusions. It would seem, nevertheless, that it was the genitalia (and perhaps the presence of the testicles) that allowed Dr Morales to reassign sex. Rather than suggesting surgery to change the sex of the individual, as may well have been more prevalent in England in cases of doubtful or disputed sex,[37] and in Spain for the most severe cases of hypospadias, seen above, a less interventionist process was advised here: changing name, clothes and probably future work activities.

Dr Morales discussed a second case in the same article. This was another case of third-degree hypospadias 'cuya deformidad congénita parece inclinar el ánimo á primera vista más á hembra que á varón' (whose congenital deformity would incline one's judgement at first sight towards [declaring] a female rather than a male).[38] A long description of Dr Morales's subject followed. As part of this discussion, he recalled the case of a colleague from Granada who had been asked to examine a woman in order to find out the cause of some abdominal discomfort. He made it known that the woman had no vagina. Upon finding this out his patient asked if any kind of operation was possible to allow for cohabitation with her husband. What the doctor's response was, we do not know. The consequences of this kind of diagnostic were immense, Morales argued. With respect to the Granada case, he asked '¿Quién sabe cuantas mujeres llamadas *marimachos* no son otra cosa que hipospádicos de último grado?' (Who knows whether many so-called tomboys are nothing more than male hypospadiacs of the most serious degree?).[39] It would seem here that gender deviance, in the form of masculinized women, was to be explained by faulty genitalia. The category *marimacho*, although with some connotation of lesbianism at a later date, was at the time primarily a gendered state. Notions of the real sex would explain away gender deviance and also, possibly, deviance of a sexual variety. But gender deviance could also explain badly formed genitalia. Medical knowledge would return the body–sex–gender alignment back to its natural course, explaining any *viragos* by illustrating how the wrong sex had been assigned to the wrong body.

Morales concluded by asserting that in humankind hermaphrodites did not in fact exist 'en la verdadera acepción de la palabra' (in the true meaning of the word).[40] While in ancient times, the hermaphrodite may have signified something monstrous, now 'las desviaciones en el desarrollo' (deviations in development) 'están sometidas á leyes fijas' (are subject to fixed laws) of organ generation. It was these laws and matters pertaining to hypospadias that midwives were ignorant of, given their rudimentary training.[41] On the subject of surgical techniques, Morales affirmed, contrary to what we have heard above, that there was no method to be employed to resolve cases of third-degree hypospadias; they were 'desgraciados seres que se quedaron en una imperfecta evolución orgánica' (unfortunate beings who have remained at an imperfect stage of organic development).[42]

The above case studies have shown the disassociation between the diagnosis of hypospadias and hermaphroditism. At times this was partial but more often than not it was a complete disassociation. It is not argued here that medical doctors, as they temporarily eliminated the category of hermaphroditism in favour of diagnosing hypospadias, necessarily ascertained less problematically or less equivocally what they understood to be the true sex of the individual. By accepting cases of hypospadias doctors were still operating from a discourse of the 'truth' in respect of the real sex of the 'patient'. What they were in fact doing was reaffirming one branch of medical knowledge, with its new surgical techniques, over another that had been questioned without remission. What the diagnosis of hypospadias allowed doctors to do was to determine sex on the basis of what they understood to be a clearer set of criteria, which were less equivocal (or tainted?) than classifications of hermaphroditism. Despite this shift, no consensus was to remain in the science of hermaphroditism for an extended period of time. Towards the mid-1910s a category that had lost currency back in the 1880s was resurrected: pseudo-hermaphroditism.

The return of pseudo-hermaphroditism

A case extensively recorded in 1913 shows how the intersecting categories of hypospadias, pseudo-hermaphroditism and the emerging predominance of the gonads in the determination of sex came to act together as combined resources. The reporting doctor, Dr F. Camacho Alejandre from the coastal town of Almuñécar, related his experience of a newly born child 'cuyos padres cortijeros de los contornos, se encuentran perplejos ante la dificultad de determinar el sexo del pequeño' (whose farm-worker parents from this district are perplexed by the difficulty of determining the sex of their little one).[43] This six-page account, complete with one photograph and one sketch, is the most extensive single case study we have located. Apart from the length of the article and the use of photographic material, common in other countries by this period but not in Spain, what is also interesting about the case is that it comes from the provinces. As in other sexual histories, 'peripheral' accounts often supply some of the richest materials. Given the importance of this material, quite some space is devoted to it in the ensuing paragraphs.

Dr Camacho recognized the urgency of defining the sex of the baby he examined in accordance with a particular set of gender prerogatives in order to successfully inscribe the infant in the Civil Register and 'saber a qué atenerse, considerándolo como lo que sea en lo que atañe a la educación y a sus relaciones familiares y sociales en el porvenir' (and to know what to prepare for, taking into account whatever [the infant] may be in terms of upbringing and its family and social relations in the future).[44]

The usual description of the infant's physical characteristics followed and the baby was found to be of a good weight and energetic. A description of the genitalia followed: two large lateral folds joined at the top housed a prepuce which partly covered an organ 'con las apariencias de un clitoris voluminoso, o de un glande pequeño' (with the appearance of a large clitoris or of a small glans).[45] Two further folds emerged from the base of the glans terminating in an orifice which appeared 'a decir verdad' (truth to tell) to be a vagina and a hymen. However, this orifice corresponded in fact to the urethra. The examination continues: on each side of the larger folds 'se nota ... dos cuerpecitos duros como dos gruesas habichuelas' (are noted ... two small bodies as hard as two large broad beans).[46] Despite their attempted manoeuvre, these 'beans' would not descend.

Such characteristics cleared up any doubt in what was a complex individualized diagnosis. The doctor explained that he was confronted with a case of perineal-scrotal hypospadias with cryptorchidism because of which the female genital organs were 'simulated'.[47] Camacho continued: 'la existencia de dos testículos ocultos y fijos ... no deja lugar a dudas, respecto al verdadero sexo del pequeño' (the existence of two hidden and fixed testicles ... allows no room for doubt as to the true sex of the little one).[48] Despite the presence of the testicles, Camacho conceded it was possible that in the depths of the pelvis a uterus and ovaries could be found but '¡es tan rara la existencia de los verdaderos hermafroditas!' (the existence of true hermaphrodites is so rare!). In previous cases, the presence of ovaries and testes in any given individual were nothing more than embryological remains and one sex always predominated, thus allowing for the individual to be classified. To use Camacho's words, in this case the 'female genital organs' were 'simulated'; they were apparently of one sex but in reality of the other.

A frank discussion then followed on the subject of the value of the medical advice contained in the report. Camacho averred that it was necessary that the parents followed his advice in order to prevent serious consequences:

> *Si estas gentes prescinden del médico en este caso dudoso, o si yo yerro mi juicio, pueden equivocar la senda y educar un niño a lo niña con lo que harán un marica o una niña a lo niño y será un marimacho.*
>
> If these people do not follow the doctor in this doubtful case, or if my judgement is incorrect, the wrong path can be taken and they may bring a boy up as a girl, thus making a fairy or bring a girl up as a boy thus making a tomboy.[49]

Such a danger seemed to respond, as in the case referred to by Dr Morales above, to anxieties over gender role rather than actual sexual orientation since it was argued later that 'Claro que luego el instinto del sexo servirá para aclarar la cuestión' (Of course, later the instinct of each sex would serve to clarify the situation).[50] The possibility of sexual inversion, however, was evoked subsequently: '¿pero no se ven muchas inversiones sexuales y precisamente estos pseudohermafroditas y criptorquidos son terreno abonado para ello?' (but are not many sexual inversions seen and it is precisely these pseudo-hermaphrodites and cryptorchids who are fertile terrain for their development?).[51] The association between hermaphroditism and sexual deviance was a long one; some doctors evidently still believed that doubtful anatomy could produce homosexual desire.

The remainder of this extensive article focused on the possible development of the baby examined and, if the testicles did not descend and develop normally, the dangers of glandular disorder, which would result in a feminine appearance. Although not the first example of an article to recognize the role of glands in human development,[52] Camacho's analysis, in addition to speaking around the contours of pseudo-hermaphroditism, hypospadias and the importance of the gonads in sex identification, also introduces the next major change in hermaphrodite science: the arrival of hormonal accounts. This new scenario is discussed later on in this chapter.

Of final note is the evident up-to-date knowledge of the question of hypospadias displayed by Dr Camacho. Much of his article was dedicated to the embryological explanations of hypospadias and the primarily degenerationist account of cryptorchidism. On these subjects Duval was referred to, as was Dr M. Katenstein, the Berlin physician cited in other pieces in preceding years.[53] In this way, Camacho's analysis forms a kind of half-way post between a predominant focus on hypospadias and the emergence of a new paradigm for explaining doubtful sex – pseudo-hermaphroditism. While, as we have said, the use of the term to designate cases found in Spain only returned in the mid-1910s, there was ample reference to the category in foreign countries in the immediately preceding years.[54]

One case in particular that was noted in the Spanish medical press is worth a short discussion. A German case reported in the *Deutscher Zeitschrift für Chirurgie* in 1909 by Dr Meixner described an example of external female pseudo-hermaphrodism.[55] A deceased eight-and-a-half-month baby who appeared to be a well-constituted baby boy was described. The 'virile member' of 2cm and lack of testes in the scrotum could have led to the diagnosis of a 'simple case' of cryptorchidism. However, the autopsy showed ovaries, a uterus and all the female organs. A histological analysis confirmed the glands to be ovaries. Drawing on Klebs, this subject was deemed to be an external female pseudo-hermaphrodite as the two glands that determined the sex were the same (both ovaries) and since they were both female glands. The external aspect was masculine, from the look of the genitalia, but the subject was indeed female, given the above data. Neugebauer was mentioned in order to confirm that this case was not exceptional.

This example gives us a taste of the changes in hermaphrodite science that were emerging. While the designation of pseudo-hermaphroditism implied a confirmation that hermaphroditism could not really exist it also entailed a reconfiguration of the field different from that which was implied by the identification of hypospadias. If diagnoses of hypospadias shifted the focus away from hermaphroditism, that is, the combination of traits of both sexes in one individual, they only did so for a while. Hypospadias implied that the genitalia spoke the truth of the individual. But when the importance of the gonads became consolidated this primacy of the genitals and the condition of hypospadias were made problematic and revised in the face of the new category of pseudo-hermaphroditism. The case discussed by Campacho could easily, it was noted, have been classified as a simple example of cryptorchidism. The newer designation of pseudo-hermaphroditism, basing itself on the glands alone and not the appearance of the genitalia seemed a more solid terrain on which to establish sex, even though Dr Campacho still admitted the possibility that he could be wrong, thus resulting in the creation of a 'fairy' or 'tomboy'. But even this new set of criteria had its pitfalls: it was not possible to examine the internal organs of individuals to be sure of their sex, by means of a procedure called laparotomy, until the mid-1910s.[56] This meant that the true sex of the individual could only often be determined in the cases of deceased individuals, such as the baby described above.

In order to understand this shift we need to examine the thought of Theodore Klebs. It will be recalled that Klebs did allow for a small number of cases to be defined under the category of 'true hermaphroditism', even though these would be very rare. True hermaphrodites were sorted into bilateral, unilateral and lateral varieties. Most cases of 'hermaphroditism', however, were not real but apparent. For these more common eventualities, Klebs described 'masculine pseudo-hermaphroditism' where there was a presence of testicles and evident development of the female parts, and 'feminine pseudo-hermaphroditism' where the ovaries were present along with some predominance of the masculine genital parts.[57] The picture was somewhat more complicated, however, than these categorizations. As cases reviewed seemed not to fit into established categories, newer, more complex ones were continually invented. In the light of these complications Klebs further divided pseudo-hermaphroditism into three subtypes.

The first of these was designated 'internal'. In the masculine internal pseudo-hermaphrodite there would be testes, the presence of a prostate with a 'masculine uterus' and the subject would be masculine externally but with some rudimentary internal feminine organs. The female internal pseudo-hermaphrodite would possess ovaries, would be internally and externally feminine with some rudimentary internal masculine organs. The second type, the internal and external pseudo-hermaphrodite, would, in its male form, possess testes, a masculine uterus and feminine genitalia and would be mainly female externally except for the testes. In the female form, the pseudo-hermaphrodite would possess ovaries, partly exterior, and internal male genital apparatus and would be mainly male externally except for the ovaries. The final category, external pseudo-hermaphroditism, would, in the male case, possess testes, somewhat feminine external genitalia, be mainly feminine externally and masculine internally. The female variety would possess ovaries, would have external female genitalia with some external masculine characteristics and would internally be mainly feminine but with some masculine features. This extraordinarily complex set of diagnostics was criticized by Pozzi as harbouring confusion over to what the adjectives 'masculine', 'feminine', 'external' and 'internal' actually applied. To quote Dreger: 'The adjectives "masculine" or "feminine" referred to the nature of the gonads (testicular or ovarian, respectively). The adjectives "internal" or "external" referred to the site of the sexual anomaly (anomalous in comparison with the gonads, which were always assumed to be the norm).'[58] It was the gonads that would prove the sex of the individual, despite the presence of other sexual characteristics such as a 'masculine uterus' or an external female look.

As was the case with any new paradigm, the thought of Klebs was not established overnight. We saw in the previous chapter that although 1882 appears to mark the arrival of Klebs in Spain his model was accepted slowly.[59] In fact, it was only really in the 1920s that Kleb's thought became more common, although, as Dreger has argued, the importance of the gonads as a sure sign of sexual identity may have preceded his system of classification. We have argued that the mid-1910s saw one of many renovations of hermaphrodite science in Spain, reconfiguring discourse on hypospadias in order to make way for a steady rise in influence of gonadal theories.

It is difficult to deny that foreign sources drove paradigm changes in hermaphrodite science in Spain. Before the influx of cases considered under the title of hypospadias there was considerable reporting and discussion in Spanish medical journals on cases of hypospadias and their treatment in the medical institutions of Paris, Berlin and Vienna. It is also possible that changes in surgical techniques in Spain spurred on this attention to hypospadias. A similar process can be detected in respect of the category 'pseudo-hermaphroditism'. While this category had been employed back in the 1870s,[60] it was from the mid-1910s and especially the 1920s that cases of 'hermaphroditism' began to be classed as examples of 'pseudo-hermaphroditism' on the basis of the gonads. This was no hegemonic process, however, and we still see a variety of diagnoses adopted, including 'hermaphroditism', hypospadias and pseudo-hermaphroditism itself. The next section traces how cases of hermaphrodites reported on in medical journals followed this new trajectory.

The influential *Siglo Médico* reported in 1916 on certain debates taking place in Germany with respect to the civil status of hermaphrodites.⁶¹ It was reported here, in a discussion on the civil status of hermaphrodites, that the 'general opinion' on registering the sex of these individuals as men or women 'más que á la conformación de los órganos genitales externos ... que se atienda a la existencia de testículos ó de ovarios' (owed more to the existence of testicles or ovaries ... rather than to the conformation of the external genital organs). But one medical figure, E. Wilhelm, believed that the general aspect of the subject and an examination of the will or desires of the person were more important than the presence of the gonads. The scenario could arise whereby 'á un sujeto que tiene glándulas seminales deba considerársele como del sexo femenino, y viceversa' (a subject who has seminal glands should be considered as of the female sex and vice versa [i.e. a person with ovaries, presumably, could be considered a male]). It was further argued that the law should be modified in order to allow for subjects to be registered not as males or females but as 'indeterminates' and, possibly, once adults to be classified according to the sex that they correspond to. Such an exchange shows how the gonadal model was not hegemonic in a country with much original discussion and theories of classification of hermaphrodites, although it was clearly a paradigm that created debate. Such a stance also shows how uneven thought on the subject was; Wilhelm's position has more in common with early nineteenth-century authors such as Marc (recall his category of the 'epicene').

What cannot be doubted is that, as in the late nineteenth century, increasingly sophisticated notions of hermaphroditism entered Spanish medical knowledge in the 1920s. A case of 'human bilateral hermaphrodism with bisexual glands (gonads)' was reported from *Gynécologie et Obstétrique* from February 1920 in the Spanish journal *La Medicina Ibera*. This individual was described as possessing an ovary and a testicle and was presented by E. Brian, Lacassagne and Lagouffe. The category of bilateral hermaphroditism corresponded to one of Klebs's categories of 'true hermaphroditism'.⁶² A further example from the French *Journal d'Urologie* recorded the case of true glandular (gonadal) hermaphroditism in an individual of 10 years. Here, there were mixed gonads with no male or female predominance.⁶³

The possibility of real hermaphroditism was, as part of the same process, narrowed down to a very small number of cases. Accordingly, from 1920 Spanish journals began to report more often on cases of pseudo-hermaphroditism. Our first example is taken from the pages of *La Medicina Ibera* and is discussed by Dr Salvador Pascual. The case study is interesting, not only because of the kinds of criteria utilized to define the individual in question, but also because it is titled 'A case of hermaphrodism'. It was immediately to be clarified as a case of pseudo-hermaphroditism. This persistence of older categories shows how 'hermaphroditism' as a ghost category never really fades away in studies such as these. It is a device constantly called upon, often to be instantly disavowed, although later, as we shall see, to be reinstated.

Dr Pascual, reporting on a case first seen in the urology clinic of Dr Mollá to an audience in the Royal Academy of Science, described in usual fashion the physical characteristics of the person in question.[64] That cases such as this were discussed in such august audiences shows once again that the subject was by no means a minor one. Indeed, as the 1920s wore on it would become a question of greater urgency to be resolved by medical doctors, reflecting the kind of interest that was encountered in cases of hermaphroditism in the 1870s and 1880s in Spain. Dr Pascual described a situation which was similar to those discussed in the works of Pedro Mata and other medico-legal doctors during the last half of the nineteenth century: a young married couple came to Dr Mollá 'con la pretension de que fuera reconocida la mujer' (with the desire that the woman be examined). This woman, the account continued, had never menstruated but her sexual desire was focused on the male sex.

There was some doubt over the 'erectile organ', 'que no tiene forma determinada demostrativa de que sea clítoris o pene' (which cannot clearly be determined as either a clitoris or a penis). Above the erectile organ there are two large labia-like folds, 'like a vulva' and in these there lie 'dos abultamientos, uno a cada lado' (two forms, one on each side). It is decided that 'Por el tacto, palpación y exploración, los órganos genitales de esta mujer corresponden al sexo masculino' (By means of touch, palpation and exploration, the genital organs of this woman correspond to those of the male sex). Furthermore, 'Desde luego, no es un hermafrodita; es un pseudo-hermafrodita' (Of course, it [*sic*] is not a hermaphrodite, but a pseudo-hermaphrodite).

What followed was an indicative discussion of the kinds of framework that Spanish doctors were familiar with at the time. Badly spelt versions of Klebs ('Rlebs') and Pozzi ('Posse') preceded these systems of classification. Following Klebs, the individual would be classified as a 'masculine pseudo-hermaphrodite', but we are not told the subtype. In terms of Pozzi's system, the individual would be classed as an 'androgynoid', that is, a man with the appearance of a woman. While both these authors' systems relied principally on the gonads to define sex, in this presentation the characteristics to be viewed in order 'to diagnose hermaphroditism' were much more diffuse. These included the general aspect, the comportment of the individual and the likes and inclinations held towards the female and male sex. All these categories, we are told, are of 'scant value' in the woman in question but, presumably, could be more indicative in other cases.

The difficulties of a reliable diagnosis were referred to in the sense that, Pascual stated, there were 'numerous cases' whereby subjects had lived their whole life as women, only to be found later, as a result of a chance occurrence or an autopsy, to be really men. In these men, desire for the masculine sex was often 'exaggerated'. In this case, however, touch, palpation and the general look of the woman were not sufficient to determine sex: 'Ganada la confianza de esta mujer por el Dr Pascual, le confesó que se masturbaba pensando en su marido o en los mozos del pueblo' (Once Dr Pascual had gained the woman's confidence, she confessed that she masturbated while thinking about her husband or about the young men in the village). Without being able to perform any other kind of perhaps dangerous internal operation, Dr Pascual asked the woman to masturbate, the product of which was examined under the microscope and found to be sperm. While the practice of laparotomies was not particularly current, evidently the microscope was a device that could be harnessed to decide cases of doubtful sex.[65]

Dr Pascual was concerned about the legal aspects thrown up by the case. Whether this was a case whereby the marriage would be dissolved or whether there was an error of person, the husband had married not a woman but 'Fulano de Tal' (A Mister So-and-So). A further cause for concern was the eventual civil category, that is, the sex that the individual should be assigned. This question for Dr Pascual 'no está resuelto' (remains unresolved), although he cited examples of authors who favoured assigning sex on the basis of the predominance of one set of characters or another.

The case is interesting because it shows how the gonads were taken as the agents that would speak the truth of the body, either through their mere presence (the two 'forms') or through their functionality (the production of sperm) and hence the possibility of insemination and 'potency'. The matter of same-sex desire was made explicit only in the sense that the error of person had entailed a 'same-sex' marriage. Finally, in apparent ignorance of medico-legal discussions on similar cases, Dr Pascual was bemused as to what any next step might be.

A second case in the same review encountered many of the same problems. Once again at the Royal Academy of Medicine (session of 26 April 1924) Dr León Cardenal related a case of 'pseudo-hermaphroditism'.[66] What this case and the previous one show is that around the 1920s, in contrast to previous decades, we see greater importance accorded to the patient's desires – although not as something of interest in themselves but as part of a diagnostic technique. Even though the patient's desires were not taken as decisive in any diagnosis or recommendation that doctors were prepared to make, their discussion may show a greater sensitivity towards men and women who came to doctors' clinics with this kind of question.[67] This approach by hermaphrodite science represents a revision of some of the tenets of earlier surgeons, such as Marc, who warned against taking the patient at his or her word or those who argued that any such account was without value.

This woman 'que se cree tal' (who believed herself to be such), was 25 years old, was married, and practised normal sexual relations with her husband; 'En verdad es un hombre' (In fact, she is a man). She had come to the clinic because of her 'double inguinal hernia'. An extirpated body, having been thought previously to be an ovary, was identified as a testicle under the microscope. A short section on classification followed: hermaphroditism was considered to be 'true' in the case of the presence of both a testicle and an ovary, the 'glándulas genitales femeninas y maculinas, aun en estado embrionario' (female and male genital glands, even though in an embryonic state). Of these cases, Cardenal noted, only three had been found. The rest were cases of pseudo-hermaphrodites.

This example of a married woman corresponded to a case of male pseudo-hermaphroditism: 'Aunque tiene órganos genitales externos de aspecto completamente femenino, e igualmente caracteres sexuales externos' (Even though he has external genital organs of a completely female aspect and also external (female) sexual characteristics),[68] the internal female genitalia were not present; there was nothing 'simulating the vagina' and no vestiges of a uterus, prostate and seminal ducts. This was the closest we would get to an analysis drawing on Klebs's system of internal/external characteristics and it is quite clearly indebted to his system.[69]

Final aspects worthy of note in this case study are the commentaries by other doctors at the end of the account. Dr Valle Aldabalde remarked that it would be of interest to see whether in the extirpated testicle it was possible to find signs of an ovary.[70] Dr Maestre, finally, argued that the lack of sexual differentiation in the internal genitalia was a result of the action of the hormones in the body of the person. However, it was necessary to recognize that recent experiments had demonstrated that female and male hormones were essentially the same.[71] This recourse to the science of hormones would open another chapter in the science of hermaphroditism, to be discussed below.

A final example will suffice to show how the concept of pseudo-hermaphroditism had gained hold and how a new set of medical techniques – the use of the microscope in histological examinations – speeded up or at least greatly contributed to its acceptance. Three cases of pseudo-hermaphroditism were discussed from a histological perspective in 1925.[72] The authors, in their introduction to the subject, remarked that the number of cases of true hermaphroditism, that is, those individuals with 'both glands' histologically proven, did not exceed seven or eight. More abundant were individuals

> con genitales externos de un sexo e internos solamente del opuesto, los que, siguiendo la clasificación de POZZI, pueden dividirse en dos grupos: androginoides, esto es, mujeres que parecen hombres, y ginandoides, o sea hombres que parecen mujeres.

> with the external genitalia of one sex and only internal genitalia of the opposite sex. These, following the classification of POZZI, can be divided into two groups: androgynoids, that is, women who seem to be men, and gynandroids, that is men who seem to be women.

These individuals were all classed as pseudo-hermaphrodites.

The three cases referred to by the authors were all of the gynandroid type in whom testicles in different stages of atrophy were found. The first case was an apparent woman, perfectly formed, with normal mammary development and whose 'psychic state' corresponded entirely to her apparent sex. By this, the authors meant that she acted entirely as a woman and, probably, her sexual desires were directed towards men. The second case was a young girl and the third recounted the case of a woman of 28, married, with no marked traits of femininity (one wonders if the gynandroid label is, in fact, appropriate in this case), with little mammary development, no uterus, and a vagina of 'ordinary dimensions'. However, two testicles were discovered on operating for a hernia.

The case for the gonads, or 'glands' as they were often referred to in Spain, being the marker of true sex, despite all the complications such a diagnosis could entail, was repeated in a final text whose author was once again Salvador Pascual.[73] Pascual remarked:

> El estudio moderno del hermafroditismo está basado en la embriología y en la influencia que las glándulas de secreción interna tienen positivamente en la determinación del sexo.

> The modern study of hermaphroditism is based on embryology and on the influence that the glands of internal secretion have in a positive manner on the determination of sex.[74]

Although Steinach and others had experimented on homosexuals following this notion, more interesting from the medico-legal point of view, according to Pascual, were cases of pseudo-hermaphroditism as these gave rise to marriages between individuals of the same sex. It was Pozzi, Pascual continued, who had divided this category into androgynes and gynandrynes. Pascual was definitively in favour of the importance of the 'genital gland' in the determination of sex:

> El diagnóstico del sexo verdadero de un hermafrodita no puede establecerse más que por la comprobación de la glándula genital anatómicamente en el curso de una operación o de una autopsia, o fisiológicamente por la menstruación o la espermatogénesis.

> The diagnosis of the true sex of a hermaphrodite can only be established on the basis of the anatomical determination of the genital gland by means of an operation or an autopsy, or physiologically by menstruation or spermatogenesis.

No other characteristic (external appearance, customs, sexual appetite, external genital organs), it was stated, could be held to be decisive. In doubtful cases, individuals would be ascribed the sex that was dominant.

We have seen how pseudo-hermaphroditism certainly became a key resource for the classification of 'hermaphrodites' from the mid-1910s onwards on the basis of the presence of a gonadal element and for the identification of true sex. Part of the route followed by this trend was the supplanting of hypospadias as an area of interest as the 1920s wore on. But the tendency towards awarding the glands a major role, if not the definitive role in the process of identifying true sex, constituted a changed medical horizon wherein histological examinations and laparotomy became possible. This did not mean that the phenomenon of hypospadias and its relationship with hermaphroditism disappeared but it did mean that the incidence of discussion of this condition fell off considerably.[75]

The 'felices años veinte': gender blurring and the rise of the 'third sex'

Dagmar Vandebosch has remarked recently that much of the work of Gregorio Marañón responded to a desire to normalize sexuality and in the process marginalized those expressions that did not conform to what we might now term strict gender and sexual norms.[76] Stemming from the nineteenth-century process of pathologization of 'peripheral sexualities', the discourse of sexologists and psychiatrists provided taxonomies and aetiologies of dissident sexual practice in the West in a process that has now been well documented.

Feminism, new roles for women, a growing homosexual subculture and new economic arrangements heralded by modernity would all exert their effect on changing gender roles in late nineteenth- and early twentieth-century Spain. The world war of 1914–18 also served to reconfigure gender patterns across the West, in particular, and Spain was not unaffected by this process. A cluster of concerns emerged: there was increased interest in healthy maternity, the prevention of venereal disease, greater analysis of sexual behaviour, the dismantling of the traditional working-class home, more women working outside the home and, in some cases, the sidestepping of marriage.[77] By the 1920s this process seemed to have gained an impetus that bred sexual dissidence, the crumbling of boundaries between the sexes and the upsetting of a strictly gendered world. For men, it seemed that their world was being torn asunder by the 'new woman' as she abandoned the home and hearth for the factory, the café or the street.[78] Many doctors looked on aghast at these changes. The medical doctor Quintiliano Saldaña illustrated this change-about in graphic terms: 'ellas se visten con nuestros trajes, lucen masculinos tocados' (they [women] dress in our clothes, they display male attire) as the male sphere was increasingly 'invaded'.[79]

The challenge of these different ways of understanding and expressing sexuality was met with at least three further kinds of response. First, this new reality was treated as a threat and women who took on roles different from those that had traditionally been ascribed to them were vilified.[80] Medical professionals would try to argue that women were not mentally or physically equipped for a role outside of the home. The Galician pathologist Roberto Novoa Santos would argue, basing his account on the thought of Moebius, whose book had argued for female inferiority in all spheres, that women were incapable anatomically, physiologically and psychologically of undertaking such new roles.[81]

A second response would be more enlightened socially and would understand women's new role as part of the necessity to shine light on the dark recesses of sexuality, taking sexual knowledge out of the realm of Catholicism and into the modern world of sexological expertise. Such a tendency grew out of the nineteenth-century rise of sexological literature penned by concerned moral and social hygienists such as Pedro Monlau i Roca, whose *Higiene del matrimonio* ran into several editions over the later part of the century. The modernization of sex can be understood as an expression of what Foucault has identified as the four axes of discourse on sexuality from the sixteenth century onwards, which gathered strength in the mid-1800s: these would include the socialization of procreative behaviour.[82]

A third kind of response to changing gender roles in the late nineteenth and early twentieth century, rather than an attempt to shore up traditional roles or to reassign them within a modernizing sexological discourse that would retain sharp divisions between the two sexes, constituted an embracing of ambiguity, androgyny and homosexuality as an aesthetic or political undertaking. The *fin de siècle* across many European countries constituted one pole around which this tendency would be expressed.[83]

Despite Marañón's and others' attempts to argue for the natural and essential basis of sexual difference and the need to differentiate between the sexes, this discourse was also met with its counterpoint.

Resistance came from multiple sites. In 1932, the feminist Benita Asas Manterota adopted a perhaps unexpected take on Marañón's thought on sexual intermediates. In an article on 'new cultures' she praised the fact that science had been made available to all and that the full understanding of sexuality could now forge ahead without impediment. Science had enabled people to understand that in the human species there were not just two sexes: 'hay tres sexos: el masculino, el femenino y el hermafrodita ... Marañón, insigne biólogo español, nos alecciona diciéndonos que "el varón-tipo y la mujer-tipo son entes casi en absoluto fantásticos"' (there are three sexes: male, female and hermaphrodite ... Marañón, the geat Spanish scientist, teaches us that 'the typical man and woman are almost entirely fantastical entities').[84] This adoption of the thought of Marañón takes us to an examination of his ideas from the mid-1910s through to the early 1930s.

Marañón: biology, feminism and sexual differentiation

Vandebosch places Gregorio Marañón within the 'generation of 1914' of writers, scientists, critics and literary figures.[85] This generation emerged out of and contributed to the notion that for Spain to advance it had to be connected to Europe and had to engage with new scientific thought. The project of the Europeanization of Spain, as the philosopher Ortega y Gasset had advocated earlier,[86] political reformism and the desire to increase the country's scientific base was what characterized this group of intellectuals.[87] Marañón, in addition to dealing with general medical issues, questions of evolution, endocrinology and histology, was also a historian and in this respect his work represents an eclectic mixture that does not conform easily to the science/literature divide. Indeed, he believed that science and the arts 'drink from the same source'.[88] His legacy is much debated, as Glick has pointed out,[89] and it has become fashionable to deride Marañón as a reactionary; we can at least say that while he advocated, for example, the decriminalization of homosexuality,[90] he was also very much a product of his times in terms of his thought on 'correct' sexual functioning and gendered spheres. Some examples will show how Marañón's understanding of the sexes was indebted to essentialist principles of biological difference. In fact, a certain degree of biological determinism allowed for the naturalization of sexual differences in his thought.

In his 1920 conference speech 'Biología y feminismo',[91] Marañón distances himself from the 'rabid partiality' displayed by Moebius in his book on the inferiority of women. In addition, it was necessary to consider legitimate feminism from a biological standpoint, not as 'algo rígido e invariable, sino como una cosa flexible, adaptable y viva, distinta para cada caso y en todo caso subordinada a las características biológicas del país donde se va a aplicar' (something rigid and invariable, but as a flexible, adaptable and living thing, different in every case but in all cases subordinated to the biological characteristics of the country where it is to be applied).[92] In Spain, the danger was that feminist aspirations were attempting a change before the necessary educational groundwork had been done.

In addition to the anabolic (energy-saving) and catabolic (energy-spending) characteristics of women and men respectively, according to Geddes and Thompson's schema, it was the ovary and the testicle that would guide their lives, govern the metabolism and mark out the very different primary and secondary sexual functions.[93] Such differences had been inscribed into humanity from the earliest times and had not arisen, as feminists ardently claimed, from the inequalities imposed on women by men.[94] Women could only escape the 'normal laws of their sex' by becoming 'masculinized'.[95] This was a form of sexual inversion whereby the masculine intellect was combined with perfect female primary sexual characteristics.[96]

These ideas prevailed in Marañón's work right through the 1920s. In his 'New ideas on the problem of intersexuality' (1928),[97] similar core biological differences between the sexes were understood to prevail. These were transferred on to the social sphere with ease. In this essay, originally published in the influential *Revista de Occidente*, we see how, despite there existing basic differences between the sexes, a certain degree of 'confusion' between the two could occur resulting in 'intersexual' types. As we will see, Marañón's notion of intersexuality responded precisely to an *idealized* full and complete differentiation between the sexes. What occurred in reality were multiple borrowings of the characteristics of the one sex by another. This was not necessarily pathological, could not be avoided completely but, through a pedagogically oriented medical approach, could be confined to a non-dangerous level. Marañón believed that at certain historical junctures, as at the time of writing, 'en algunas mujeres adolescentes una tendencia dinámica impetuosa ... se exacerba y acerca a la muchacha al esquema viril' (in some adolescent women an impetuous dynamic tendency ... is exacerbated and approximates the girl to the male type).[98] This tendency also occurred in later life during the climacteric period when women would often become more masculine and their inclination and aptitude for political, philanthropic and artistic activities increased.[99] Women, in order to move beyond the infantile stage of their sex would aspire towards virility, which meant dynamism, something that the feminist movement embodied.

The articulation of differences between the sexes on the one hand and the very fact of their proximity and *mélange* on the other, together with the marking out of stages, or a 'chronology' of sexual development, lies at the heart of Marañón's theory of sexuality. In what follows, a number of texts from 1915 through to 1930 on the intersecting subjects of intersexuality, the chronology of the sexes and endocrinology are examined within the context of international developments on these areas. To finish this chapter, a number of case studies of hermaphroditism appearing in medical reviews are discussed in the light of Marañón's thought. It is worth saying now, however, that while Marañón was an extremely influential figure in the study of hermaphroditism, intersexuality and endocrinology, his thought did not go unchallenged. Nor did his thought imply that previous models were instantly swept away. Marañón's theory of intersexuality was yet another model that competed in explanations of what doctors from a variety of perspectives had denoted 'hermaphroditism'.

The rise of endocrinological explanations for sex

In Carmen de Burgos's novel, *Quiero vivir mi vida* (Madrid, 1931), Alfredo, a thinly veiled Marañón, explains his theories of intersexuality. The reaction of the public was outrage: 'Hubo una protesta general que se extendió por el salón. Todos querían mantener la integridad de su sexo' (There was a generalized protest that spread throughout the room. Everyone wanted to maintain the integrity of their sex.)[100] This cross-over between science and the arts would have probably been viewed favourably by Marañón himself and it suggests that ideas about sexual fluidity (and resistance to it) were in the air at the time. If the Spanish public had been fascinated by the feats of the female aviator Ruth Elder, of the first female Spanish aeroplane pilot María Bernaldo de Quirós and the monumental arrival of the train from Asturias into Madrid, driven by Señorita Careaga,[101] other publics had been fascinated by the 'rejuvenating' experiments of unorthodox medical figures such as Steinach and Voronoff. The new science of the 'internal secretions' likewise made its impact in Spain.[102]

A history of the rise of the endocrinological sciences in Spain is not the remit of this section.[103] If in the early nineteenth century the locus of femininity was understood to be the uterus,[104] in the mid to late nineteenth century this had shifted to the ovaries. The ovaries, as we have seen, became the object of surgical intervention and, along with the male gonads, the sign of true sex. In the early twentieth century, however, the locus of femaleness shifted once again and became located at least partly in the hormones, which were secreted by a number of glands.[105] Such a move enabled the delineation of boundaries between obstetrics, focusing on the uterus, and gynaecology focusing on the ovaries.[106]

Sex endocrinology was dominated in the early twentieth century by two approaches: the biological and the chemical. Up to the 1920s the biological model prevailed whereby the ovaries were thought of in terms of regulation of the nervous system. In the 1920s gynaecology first introduced the idea that the ovaries secreted chemicals. Physiologists began to use the concept.[107] In the 1910s physiologists believed that the embryo was affected by environmental and physiological conditions whereas geneticists favoured the action of the chromosomes as broadly independent of circumstantial conditions. Sex endocrinologists believed that hormones provided the missing link between the two spheres of action. Sex determination may take place by means of the chromosomes in accordance with Mendelian inheritance but sex differentiation took place in response to hormonal action.[108]

Instead of the gonads being the agents of sex differentiation, then, the hormones were understood as the chemical 'messengers' of maleness and femaleness. Scientists designated hormones as sex-specific in origin and function in the period 1905–20, that is, there were understood to be 'male' sex hormones and 'female' sex hormones which would make the embryo and the growing human being into a clear male or female.[109] According to the most enthusiastic of their proponents, hormones would govern all aspects of human development and existence. Indeed, the future Regius

Professor of Medicine at Oxford, Walter Langdon-Brown, would declare in 1923 that we 'are marionettes of our glands'.[110]

The term 'hormone' was a result of the reformulation of the doctrine of 'internal secretions' by the British physiologist Ernest H. Starling. Revising the theory of internal secretions established by C. E. Brown-Séquard (1817–94) and his assistant Arsène d'Arsonval (1851-1940),[111] Starling argued in 1905 that the 'chemical messengers or "hormones" as we may call them, have to be carried from the organ where they are produced to the organ which they affect, by means of the blood'.[112] In the early 1890s, Brown-Séquard produced testicular and other extracts, which could be used to cure diseases whose aetiology was unknown but whose pathology was associated with a particular organ tissue.[113] Subsequently, a distinction was made between the hormone, a chemical derived from animal tissues, which had specific physiological effects and an internal secretion, an entity whose absence resulted in disease, a hypothetical entity. Hormones could be isolated in the laboratory; internal secretions were implied by clinical observations.[114] In practice in Spain, the terms 'glandular secretions', 'internal secretions' and 'hormones' tended to be used interchangeably. We have seen that in Spain this new line of research did not remain unreported. In 1900 the *Gaceta Médica Catalana* announced a case of ovary transplant in the United States and in 1908 Dr Serrallach wrote up an extensive and up-to-date analysis of the action of the internal secretions.[115]

The period before 1920 was characterized in Spain by a primarily physiological view such as that of Serrallach. According to Thomas Glick, Spaniards conceived such research as part of clinical medicine under the rubric of 'organoterapia'.[116] It was Gregorio Marañón, however, who would move beyond the physiological aspects of glandular secretions in order to assess their role in the fixing and maintenance of sexual identity. Marañón was certainly the most influential figure and the lead scientist in the Institute of Medical Pathology at Madrid Hospital, reorganized in 1925 under his auspices.[117] It was from here that many of the cases of endocrinology experimentation and studies of cases of hermaphroditism would emerge.

In an account of the development of the science of endocrinology written in 1915 by Marañón three stages were identified. There was a 'pre-endocrine era', a period of latency up to 1900 and a period of explosive growth from 1900 to 1910.[118] Some years later, in 1922, in a further evaluation of the history of developments in endocrinology Marañón summarized the current state of the 'doctrine of internal secretions' and it is here that we see the relationship between sexual differentiation and the sex-specific action of the hormones made explicit.[119] In this account he discussed a period of latency, once more, during which time Brown-Séquard's doctrine experienced little uptake, a 'hyperbolic' period in which the idea became broadly disseminated and during which concepts lost their 'original composure' and were made to 'explain everything'.[120] This was followed by a reverse process, a questioning of some of the more extravagant claims and a 'crisis'. This crisis, as Glick has pointed out, was engendered by the American E. Gley's reining in of some of these excesses, which were discussed in his four lessons on endocrinology in 1919 at the Societat de Biologia (Barcelona).[121] After this period of crisis, endocrinology entered its 'classical' or mature phase, when extreme attitudes were no longer admitted.[122]

In a section of his Royal Academy of Medicine acceptance speech Marañón outlined the physiological role of the internal secretions. This was subdivided into five main areas: intra- and extra-uterine growth; the regulation of the nutritional metabolism; the 'genital function'; the connection between the nervous and endocrine systems; and the defence of the organism against toxins and attacks.[123] For Marañón, before endocrinological accounts, the pathology of the sexual functions was the subject of explanations of little more than 'literary value'; today, however, the whole physiological and pathological nature of the sexual problem could be analysed precisely.[124] Sexual functions could be broken down into two main areas, the somatic and the functional. The somatic aspect accounted for the development and morphology of the primary and secondary sexual characteristics.[125]

The functional aspect was also divided into primary and secondary functions. The primary ones referred to the 'sexual impulse' and the ability to reproduce, in both sexes, and menstruation, pregnancy, birth and lactation in women. The secondary functions included those aspects that were different in the sexes, that is, physical aptitude, the predominance of sentiment or intellect in psychological activity, and the predominance of the maternal instinct or the instinct in general in social life. These aspects were clearly delimited between the sexes, guided by the internal secretions. On a second level, the thyroid and other glands also affected these differences.[126] At the time of writing, Marañón argued, it was admitted by nearly all that the fertilized egg was bisexual in nature (that is, contained the two sexes) and that sex determination, despite this initial hermaphroditism, in either a male or in a female sense depended on the action of the hormones, generally from the testes and the ovary.[127] However, whichever sex emerged there remained

> *los caracteres del sexo contrario ... amortiguados, latentes, hasta edad bien avanzada de la vida, pudiendo, ya espontáneamente, ya en condiciones experimentales, revivir y determinar una inversión sexual más o menos acentuada.*
>
> the characteristics of the opposite sex ... hidden, latent until the later stages of life, can, either spontaneously or under experimental conditions, return to life and cause a more or less pronounced form of sexual inversion.[128]

The implications for hermaphroditism and intersexuality are clear as either or both primary and secondary sexual characteristics could stall, could be underdeveloped or could emerge at a later period. For example, so-called 'pre-pubertal pseudo-hermaphroditism' would wane once the endocrinal function of the gland developed fully; if this did not occur the physician could intervene.[129] This clinical intervention would save pubertal conditions from worsening, would intervene against 'pre-pubertal eunuchoidism', ovarian deficiencies, sexual inversion and cases of hermaphroditism.[130] There could even be a kind of fusion between the endocrine sciences and psychotherapy.[131] Such was the promise of the science of endocrinology as Marañón saw it in 1922.

Marañón's account of the doctrine of the internal secretions, however, did not take into account a discovery which would rock the basis of thought on the subject in the 1920s. One historian writes: 'In the early 1920s the dualistic idea of maleness and femaleness as clearly defined hormonal states became a topic of debate in the scientific community.'[132] If certain hormones produced in the male body were supposed to create and maintain maleness and different hormones in the female body femaleness, in 1921 the first challenge to the sex-specific nature of sex hormones arrived when the Viennese gynaecologist Ofried Fellner published an article on the growth of the uterus in female rabbits after treatment with extracts of the testes. Later, in 1927, Dutch biochemists announced that they had found the female hormone not only in the testes but also in the urine of 'normal' men.[133] This find was confirmed in 1934 when Bernhard Zondek published a piece on the 'Mass excretion of oestrogenic ("female") hormone in the urine of the stallion' in *Nature*.[134] What was to be done in the light of these discoveries, which rejected the idea of sex-specific hormones? Oudshoorn asks how those substances in male organisms that possessed properties classified as being specific to female sex hormones could be labelled: 'Scientists decided to name these substances female sex hormones, thus abandoning the criterion of exclusively sex-specific origin. Female sex hormones were no longer conceptualized as restricted to female organisms, and male sex hormones were no longer thought to be present only in males.'[135]

Rather than the 'sex antagonism' supposedly found in different hormones, as proposed by Steinach, what was now proposed was the co-operation of hormones in the development of, for example, the male secondary sexual organs such as the seminal vesicles and the prostate.[136] Marañón's analysis of endocrinology became increasingly coupled to his developing theories of the chronology of the development of the sexes and the existence of intersexual states. These ideas are discussed in the next section.

The chronology of the sexes, the intersexual states and the influence of the internal secretions

Marañón's thought on these subjects was developed throughout the 1920s and culminated in his 1930 work *La evolución de la sexualidad y los estados intersexuales*.[137] This extraordinary volume, which charts the development of the sexes and their primary and secondary differences in extensive detail, also discussed hermaphroditism, hypospadias and cryptorchidism. Marañón had stated in his 1927 set of conferences given in Havana (later published by the *Siglo Médico* in 1928) that hermaphroditism, pseudo-hermaphroditism, 'bisexuality of the instinct' and homosexuality had been present in humanity since its beginnings.[138] Indeed, hermaphroditism and homosexuality 'no eran sino manifestaciones del mismo fenómeno' (were merely manifestations of the same phenomenon).[139] In fact, endocrinological studies, more than anything else, had shown that homosexuality was a product not of a psychological perversion or crime but simply the consequence of the presence of the two types of glands in the same individual.[140] Instead, however, of two extreme poles of pure virility and pure femininity there was 'una serie inacabada de estados intermedios que bordean la frontera imprecisa que separa la normalidad de la patología' (an unfinished series of intermediate states that occupy the imprecise frontier between normality and pathology).[141] A major inspiration for Marañón's theories was the work of Richard Goldschmidt, the 'American zoologist'.[142] In his 1917 article, Goldschmidt opposed the use of the term 'sex-intergrades', replacing it with 'intersexe', 'intersexual' and 'intersexuality' because these 'terms can be used in all scientific languages'.[143] By means of a review of experiments on a variety of animals and insects, Goldschmidt argued that endocrine factors were primary in determining sex differences. His book *Mechanismus und Psychologie der Geschlechtsbestimmung* further elaborated on this idea and was published in 1920. Marañón acknowledged, however, that Goldschmidt's use of the term 'intersexuality' was more limited than his own.[144]

While Marañón did not believe that the intersexual states were necessarily pathological, they were clearly to his mind undesirable. In order to prevent the dangers of infantile and adolescent intersexuality, that is, virilism in girls and, more commonly, effeminacy in boys, a pedagogical programme was called for. The route this should take was clear: '*no hay otra educación que la diferenciación sexual*'.[145] Sexual ambiguity was, as Marañón had said, part and parcel of the human legacy. But sexual differentiation increased as people got older and as humans became more 'civilized'.[146] In civilized stages, however, the 'espina de la intersexualidad' (thorn of intersexuality) remained in humanity's side. It was necessary, therefore, to 'extinguir los restos heterosexuales, ayudar a la naturaleza en su tarea de destruir la intersexualidad' (extinguish the heterosexual remains, to help nature in her task of destroying intersexuality).[147] How would this be achieved, given the fact that intersexuality was based in an organic state? Any child, with manifestations of intersexuality, 'por leves que sean' (however slight they may be), should be submitted to 'un tratamiento farmacológico opoterápico apropiado' (an appropriate form of pharmacological hormonal-therapeutic treatment).[148] Such chemical treatment, however, would be secondary to more general and psychological methods, 'dirigiendo con una disciplina severa los instintos ya torcidos hacia su meta normal' (directing with severe discipline the already twisted instincts towards their normal aim).[149]

The idea of the embryo beginning as a 'bisexual' entity, which steadily, under the influence of the glands, became differentiated into one sex or the other was elaborated upon by Marañón throughout the 1920s. In addition to the idea that traces of both sexes in one individual did not entirely disappear,[150] Marañón understood the sexes in terms of a hierarchical developmental order. Women were an imperfect version of the male but tended to develop throughout their lives towards virilism or masculinity, especially at the time of their sexual 'crisis', the climacteric. This was seen as positive and signified greater attainment in the physical and intellectual sphere.[151] If, however, the man slipped back into a more feminine form or mentality, this was understood to be regressive. Despite this, an ambiguous pubescent phase in boys was natural but had to be carefully managed.[152] Such a framework allowed Marañón to divide intersexualities into two main areas: the permanent forms of intersexuality such as male and female hermaphroditism and pseudo-hermaphroditism, cryptorchidism, hypospadias, gynaecomastia, feminization, homosexuality and all forms of sexual inversion; 'transitory' intersexualities included the masculinizing tendencies of women who were pregnant.[153]

His 1930 book, *Evolución de los estados sexuales*, contained two relatively short chapters which applied these kinds of ideas to this form of intersexuality. In a previous chapter in this same volume Marañón had stated that, given the bisexual nature of all organisms, and most evidently of hermaphrodites, he refused to classify the latter into 'true' and 'false' hermaphrodites. Having said that, a 'merely clinical' exigency aided a division into 'hermaphroditism', where the gonads ('órganos germinales', germinal organs) were true ovo-testes, which affected the totality of primary and secondary characters. A second division, for convenience's sake alone, would speak of 'pseudo-hermaphroditism', whereby the sexual organs would apparently conserve their sexual differentiation but in which case the rest of the sexual characters may share elements of one sex or the other. This classification was, in this sense at least, strikingly close to traditional systems.[154]

Given these distinctions, the chapter on 'hermaphroditism' defined the condition as '*aquel estado en que coexisten ambos tejidos germinales en una misma gónada (ovario-testes), con intersexualidad del resto de los caracteres sexuales*' (*that state in which both germinal tissues coexist in the same gonad (ovo-testes), with intersexuality in the remainder of the sexual characters*).[155] These cases would be extremely rare, 'true monstrosities', but Neugebauer, Venturi and Reifferscheid, amongst others, had recorded a small number of examples in humans. Following Klebs, but without naming him, the ovo-testes could be, according to Marañón, bilateral or unilateral, that is, they could exist on both sides of the body or on one only. The 'alternate' hermaphrodite possessed one testis on one side and an ovary on the other. Whichever way, an 'investigación *minuciosa y total de las dos glándulas del individuo*' (*minute and complete investigation of the two glands of the individual*) was necessary,[156] but this was only possible during an autopsy or as a result of some extraordinary surgical intervention. The dangers of making decisions on the sex of the individual were illustrated by a case documented by Salen in Berlin in 1899. If only the left gonad had been extirpated it would have been seen to be an ordinary ovary. The mixed gland was located on the right. The diagnosis would have been different depending on the surgical practice.

Marañón was convinced, in the light of this kind of reasoning, that 'true' hermaphrodites were more common than had been thought. Most cases did not undergo a histological analysis of the gonads but instead a clinical evaluation.[157] But the situation was further complicated by the fact that even if the two gonads had been examined, the finding of one sexual tissue was not a guarantee that no histological elements of the other sex were present. There were, then, four explanations for the existence of hermaphroditism. The first, following Krabbe, proposed that other-sex traces were found outside of the gonad, such as in the uro-genital tract. Secondly, following the recent discoveries by Zawadowski and Lipschütz, an apparently normal gonad could indeed secrete substances of both sexes, as a kind of throw-back to its undifferentiated or bisexual stage.[158] Thirdly, referring to other bodily functions, Marañón noted that researchers such as Frank and his school had found the female hormone in men's blood, suggesting a degree of intersexuality.[159] Fourth, in those cases in which the sex is of one type and the gonad is of the other, it was possible that the gonad was *originally mixed* (that is, was an ovo-testis) and that the tissues of one sex had atrophied. But the morphological sexual characters of the 'first' sex would be indelibly expressed on the body.[160] Finally, in accordance with his theory of the chronology of the sexes, by his own admission only scantly recognized in treatises on the subject such as that of

Neugebauer, Marañón believed that the morphology of the hermaphrodite, as in both sexes in 'normal' subjects, responded to a process of evolution over a period of time. Given that there were two sexes present, the subject would follow the path of both male – in early life, approximating to the female sex – and female – in later life, approximating to the male sex.[161] Despite this general trend, there would be no huge difference between the two sexes in the hermaphrodite, unlike in 'normal' persons.[162]

Marañón's schema evidently problematized hermaphrodite studies to date. His work confirmed the trend away from taking the gonads as a sure sign of one sex or the other, or fixing sex in accordance with the predominance of one sex. Rather, his system complicated classification by arguing that hermaphroditism was just one (extreme) form of intersexuality and that intersexuality was a state that merely arose because of the 'bisexual' nature of the embryo, glands and, ultimately, the action of the hormones. The figure of the hermaphrodite seemed to be omnipresent, flooding the gradations between one sex and the other, but also absent as a category so broad and commonplace that it dissolved itself in a sea of sexual intermediates.

The next chapter in *Evolución de los estados sexuales* focused on pseudo-hermaphroditism. The big difference between hermaphroditism and pseudo-hermaphroditism in Marañón's mind was, as we have noted, the question of the bisexual or mixed nature of the gonad. In the case of pseudo-hermaphroditism, there was not, at least apparently, a mixed gonad. The sexual characters, however, could well be bisexual. Included in this bracket were 'individuos con apariencia femenina y testículos' (individuals with a feminine appearance and testicles) and 'individuos con apariencia masculina y ovarios' (individuals with a masculine appearance and ovaries).[163]

The category of pseudo-hermaphrodite in this schema, however, was a notional one. In fact, Marañón was to argue that pseudo-hermaphrodites did also share this bisexual gonadal nature but the latter just had not yet been discovered. There were several possible reasons for this: because a part rather than the totality of the gonad had been examined; because the tissue of the other sex resided outside of the gonad; because the bisexual aspects were found in the interstitial tissues rather than in the germinal tissue, meaning that a single-sex ('monosexual' in Marañón's words) appearance prevailed but a bisexual function remained; because a gland that now appeared to be single-sex could have been bisexual in a former period of evolution.[164] Of course, such a wide range of explanatory criteria in fact demolished the class of pseudo-hermaphrodite. In this sense, it would conform to Marañón's aim of refusing to distinguish between real and false hermaphroditism. A final explanation was that other glands of internal secretion, such as the hypophysis could have a direct influence on the emergence of secondary sexual characters.[165] Certain cases of pseudo-hermaphroditism, such as those of virile appearance and female glands, could arise from the malfunctioning of one of these glands or, in the case discussed, the pathological hypertrophy of virilizing glands. Nevertheless, according to classical accounts, there were two types of pseudo-hermaphroditism: the traditional category of the masculine pseudo-hermaphrodite, the androgyne, a man who looks like a woman, and the feminine pseudo-hermaphrodite, the gynandryne, a woman who looks like a man. These denominations, however, should be rejected, according to Marañón, since they could allow for an equivocal idea as to the 'verdadero sexo' (true sex) of the individual to prevail. For this reason, Marañón favoured the terms 'varón pseudohermafrodita' (man pseudo-hermaphrodite) and 'mujer pseudohermafrodita' (woman pseudo-hermaphrodite).[166]

The attempt to further distil and refine the classification of hermaphrodites was, as we have seen, prevalent from the early nineteenth century and Marañón's schema was the latest in a long line. Such attempts, however, always ended up falling back on themselves in order to justify their distinctions. Despite not accepting the division 'true' and 'false', for Marañón the pseudo-hermaphrodite was 'un ser neutro, equívoco aunque en menor grado que en el hermafrodita verdadero' (a neutral, equivocal being, although to a lesser extent than the true hermaphrodite).[167] It was only a histological analysis that could differentiate the two types.

Both, nevertheless, were subject to the same kind of chronological and developmental process as was outlined above in the case of the 'hermaphrodite'. In this sense, the androgyne instead of undergoing a brief feminine phase in puberty, as in 'normal' men, has a long and intense femininity and an unstable virility.[168] Again, the true sex of these individuals was to be determined by clinical analysis. Marañón cites the case of a Swiss national examined by him who had been considered a woman up to the age of 25 despite certain virile characteristics. She had not experienced menstruation but married, split up from her husband after some years, developed a 'voz profunda, barba y bigote, vello en el tronco, gran corpulencia general' (deep voice, beard and moustache, body hair, general large corpulence). These secondary characteristics, we presume, because Marañón does not tell us, were to be taken as sign not only of her intersexuality (pseudo-hermaphroditism) but of her manhood.

A final comment should be made with regard to the status of hypospadias and cryptorchidism in Marañón's theories before examining the reporting of a number of case studies. In contrast to what appeared to be a decoupling of these two phenomena from hermaphroditism up to at least the mid-1910s, Marañón asserted that both hypospadias and cryptorchidism accompanied more general signs of femininity and that they constituted '*verdaderos estados intersexuales*' (*true intersexual states*) in contrast to what some authors had argued.[169] This is not surprising as such; Marañón's extremely broad range of intersexual characteristics covered a wide variety of apparent or 'truly' mixed sexual characters.

The application of gonadal, hormonal and intersexual criteria in cases of 'hermaphroditism', 1928–1930

During this period clinical reports of suspected hermaphrodites drew on a variety of resources in order to identify the kind of hermaphrodite doctors examined and the true sex of the individual. The category of pseudo-hermaphroditism had gained currency from the mid-1910s, supplanting a period of relative silence on the subject when other medical conditions such as hypospadias drew more attention. The renewed interest in pseudo-hermaphroditism did not, at first, signal a return of the real hermaphrodite as s/he was dismissed as impossible or near impossible. However, with the rise of endocrinological explanations of sex, the physical presence of the gonad was deemed less important than the *function* this gland (and other glands) might have in determining and maintaining sex. In addition, the definition of hormones and their function in Spain continued to attract attention during the period when Marañón was writing and analysing cases. On the one hand, it was acknowledged that the use of the term 'female sexual hormone' was 'amplio e indeterminado' (broad and imprecise), covering many different things.[170] On the other, it was argued by Dr Torre Blanco that 'No son, pues, las glándulas sexuales (testículo y ovario) las encargadas de determinar el sexo: su misión, a la que eficazmente coadyuva todo el sistema endocrino, es *mantener* la diferenciación sexual ya establecida desde el momento mismo de la fecundación' (It is not, however, the sexual glands (testicles and ovaries) that are in charge of determining sex. Their mission, which is efficiently supported by the endocrinal system, is to maintain sexual differentiation established at the very moment of conception.)[171] Any disruption of this system would entail 'formas de intersexualidad y de inversión bien explicables' (forms of intersexuality and inversion that can be explained).[172] Such action would also explain 'la semejanza que entre ambos sexos existe en las fases de la vida en que las glándulas sexuales no han despertado aún de su letargo infantil o han dormido ya con el sueño propio de la vejez' (the similarity that exists between the two sexes in those phases of the lifespan when the sex glands have not yet awakened, the infantile stage or when they have declined in the sleep of old age).[173] So, during this period, we see a number of competing accounts to explain hermaphroditism, from the gonadal model through to the endocrinological.

The first case we review is one we have come across earlier: the 'hermaphrodite' that was exhibited at Madrid fairs. In the report, Marañón reaffirmed his idea of the numerous categories that made up the class of intersexuals and repeated that the existence of real hermaphrodites was probably more extensive than previously thought. This would be the case of 'un intersexual que se exhibe en las ferias de España y que ha sido recientemente estudiado en nuestra clínica' (an intersexual who is exhibited in fairs around Spain and who has recently been studied in our clinic).[174] He or she was 60, had menstruated, and the distribution of body hair, fat and mammary glands were as of a woman. The voice and head hair were virile and 'frecuentes relaciones sexuales con mujeres' (frequent sexual relations with women) had taken place. There was a small penis, equivalent to that of a boy's of 15 with no urethra. This opened 2cm below the penis and there was a small vaginal conduct. The large labia 'tienen el aspecto de bolsas escrotales vacías' (looked like empty scrotal sacs). The diagnosis was made in accordance with Marañón's theory of intersexuality (this would be a 'hermaphroditic type of intersexuality') but the precise nature of the individual, whether s/he was a real or pseudo-hermaphrodite, could only be confirmed by a histological analysis, something impossible to carry out at present. What seems to be most important here is not the true sex of the individual but the *class* of hermaphrodite (read intersexual) s/he may be. If the endocrinological developments of the 1920s with their succession of hypotheses 'illustrates how the dualistic conceptualization of sex was gradually abandoned',[175] Marañón's analysis suggests that the category of male and female were less definable and perhaps less sought after by some clinicians, especially those under his influence.

Not every case discussed in this period followed precisely this paradigm. In an extensive report, the gynaecologist M. Usandizaga and the pathologist G. Sánchez Lucas described a case of 'masculine hermaphroditism' in 1931.[176] 'M.O.' of 63 years, a widow, remembered having felt two nut-like nodules in her groin since she was small. From the evidence given it would seem she attended the clinic because of a double hernia. She dressed like a woman and had formerly attempted sexual intercourse with her husband but had stopped doing so. She said she had always experienced inclination towards the male sex. She possessed 'typical' genitalia and a clitoris of 3cm. Below this there was a 'very shallow canal' that terminated in a vaginal opening. No uterus could be detected after vaginal and rectal examination (under anaesthetic). Dr Usandizaga performed the hernia operation and extracted a testicle from each side.[177] Macroscopically these were declared 'normal' and, despite some comments on the abnormality of tissue and seminal ducts, microscopically these did not present a departure from normal testes. After the operation, the 'enferma' (female patient) was found to be progressing well. However, because of complications arising from the surgery, the woman died a month afterwards; a full autopsy was performed and nothing was found to be particularly out of the ordinary.

The rest of the report focused on current theories of hermaphroditism. These were as eclectic as the report was long. Some elements merit our examination. The authors declared that there was no unanimity in terms of distinguishing true from pseudo-hermaphrodites. Neugebauer was thought to be the strictest, reserving the category of hermaphrodites for those that could procreate successfully. Most other authors required the presence of both 'genital glands' rather than their procreative functioning in order to declare true hermaphroditism. Reflecting the changes in the field the doctors acknowledged that 'en los últimos tiempos se tiende a ampliar todavía más el concepto de hermafroditismo' (recently there has been a tendency to expand the concept of hermaphroditism).[178] Hoepke, for example, defined hermaphroditism as any congenital alteration of the organism that brought together male and female characteristics in the same body. This stance would confirm the disappearance of the term pseudohermaphroditism and could be compared to Marañón's elaboration of the notion of the intersexual states. While Neugebauer had remarked that few real hermaphrodites existed, both Hoepke and Marañón would suggest that they 'son muchos, tantos' (are many, so many) and that the archetypal male and female were nothing more than fantasies.[179] Hoepke's system of classification was then recalled. There were five types: germinal or glandular hermaphroditism; tubular, where the external genital organs were normal but the conducts to them were of the other sex; genital, where the genitalia did not conform to the 'character of the germinal gland'; somatic; and, psychic.[180] In the case discussed above, the subject possessed 'genitales externos ... completamente femeninos' (completely female external genitalia), a normal-sized clitoris and a small vagina. But male genital glands were also present; there was evidence of cryptorchidism. No ovarian tissue was found. On this basis, the case was defined as a 'male genital somatic hermaphrodite'. This complex diagnosis even went beyond that of Hoepke, combining two of his classes in one.

Of additional interest are the possible mechanisms given for the production of hermaphroditism. Three 'indisputable' facts were already known: genital malformations arose as a quantitative difference in the development of organs that existed normally in the embryo phase or in adulthood – they did not appear from extraneous causes; all humans passed through a bisexual stage with one sex eventually prevailing over the other; while in certain animal species real hermaphroditism did exist, in humans there was a *mixture* of somatic and psychic elements rather than the presence of both sexes.[181] There followed an extensive discussion of the possible pathogenic mechanisms that caused hermaphroditism. There were six possible causes, all variants on cell structure and the influence of the glands. The last cause was the one favoured:

> *La persistencia de la fase bisexuada en las glándulas genitales ... la mezcla de elementos glandulares de ambos sexos en un mismo individuo determinaría la aparación [sic] de los caracteres sexuales también de los dos sexos.*[182]

The persistence of the bisexual phase of the genital glands ... the mixture of glandular elements of both sexes in one individual, would determine the appearance of the sexual characters of both sexes.

Marañón, it was noted, adhered to this last explanation. In those cases where a bisexual gland had not in fact been located there were four explanations. First, there was the possibility of the existence of 'restos glandulares aberrantes del otro sexo' (aberrant glandular remains of the other sex) in other parts of the body and genitalia. Secondly, the interstitial cells would take on the role of the glands in respect of the production of hormones. There would be male interstitial cells in the ovary and female interstitial cells in the testicle. Thirdly, although there was no histological difference to be found in the gland, it could recover the ability to secrete hormones of both types, as it had done originally. Fourth, there existed the possibility that the gonad was originally mixed and one of the sexed components atrophied leaving 'los caracteres sexuales correspondientes de una manera indeleble' (the corresponding sexual characters indelibly).[183] The authors argued against the existence of truly bisexual glands with tissues of both sexes in favour of a functionalist approach whereby a 'persistence' or an 'exaggeration' in the growth of organs that exist normally in the embryonic period would combine factors of both sexes. This notion of excess put an endocrinological gloss on Aristotelian thought on the hermaphrodite as subject to extra growth or supplementary parts because of disequilibrium in the humours.

A highly endocrinological set of explanations such as that used for the hermaphroditic condition of 'M.O' was not present in the report that the surgeon Dr Cardenal presented to the National Academy of Science in March 1931.[184] While the deciding factor for the sex of 'una enferma, muchacha de quince años de edad' (a female patient, a girl of 15 years of age) was the discovery of two testicles in the course of a hernia operation, there was a histological report but no reference to hormones. Furthermore, the category of 'pseudo-hermaphroditism' was retained. Another question reared its head, however. Cardenal asked: '¿Debe engañarse al sujeto respecto a su verdadero sexo, o bien, en este caso particular, debemos dejarlo que se le siga creyendo mujer y que haga su vida como tal?' (Should the subject be deceived as to her true sex, that is, in this particular case, should we allow her to continue to believe that she is a woman and to lead her life as such?) This was one of the few occasions that physicians stopped to debate this point. Usually, the genitalia or the gonads were read as a sure sign of sex that was to be implemented in the person concerned.

Marañón intervened in the debate to press for his proposal to abolish the distinction between true and pseudo-hermaphroditism and to encapsulate all such cases under the rubric of 'intersexuality'. We see an increasingly individualized approach adopted. Marañón argues: 'Respecto al problema planteado por el doctor Cardenal acerca de la conducta a seguir con estos sujetos, la solución depende ... de las circunstancias individuales.' (With respect to the problem brought up by Dr Cardenal on how to proceed with these subjects, the solution depends on ... their individual circumstances.)[185] In this example, the hormonal explanation was clearly seen as a new but disputable key to determining sex. Dr Gimeno opined that 'el problema de los estados intersexuales se ha llegado a ahondar de un modo excesivo' (the problem of the intersexual states has been plumbed in an excessive manner) and that 'varios hombres de ciencia ... dicen que incluso el testículo puede obrar como ovario' (various scientists ... say that even the testicle can act as an ovary).[186] This resistance to the new doctrine of mobility of the sexes and revision of the dualistic nature of both the sexes and hormones was repulsed by Marañón who referred his colleagues to Frank, who had found ovarian follicular hormone, the female hormone, in the blood and urine of stallions.[187]

Despite the fact that Marañón, in his *Evolución de los estados intersexuales* (1930) and in declarations such as that above, appeared confident in his assertions on the action of the endocrine glands in the development of sex, his clinical approach may not have been so secure. In fact, while he may have discussed virilization in women with a high degree of certainty, together with its hormonal treatment, a discussion on 'histological intersexuality' and 'chemical intersexuality' in the same year appears to be placed on the cusp of two explanatory paradigms, beset with a number of doubts.[188] Marañón discussed the gonadal explanation of intersexuality within his framework of the existence of the two glands and the atrophy of one of them in cases of hermaphroditism. Such was also the case in examples of late virilism in women and had been proved histologically. This had not been proven, however, in many other cases and in cases of pseudo-hermaphroditism.[189] Recently, Marañón notes, histological criteria have been replaced by chemical ones in order to determine intersexuality. A possible cause of error in diagnosis was that gonadal tissue was able to secrete substances of the other sex, a phenomenon proven by Lipschütz. In his own research, Marañón had attempted to find the female hormone in thirteen cases of intersexual men. This sample was composed of 'tres homosexuales sin el menor carácter de intersexualidad; en uno fué positiva la hormona femenina – foliculina – en sangre y orina, siendo positiva; en el restante fué negativa su presencia en orina y sangre' (three homosexuals without the slightest trace of intersexuality; in one of them there was a positive presence of the female hormone in blood and urine; in only one it was found in urine, being positive; in the remainder there was negative presence in blood and urine).[190]

Despite this rather poor set of results, Marañón could optimistically report that in the majority of cases of such men reports had been positive for the presence of the female hormone, thus proving his intersexual thesis. It could be objected, Marañón conceded, that the mere presence of the female hormone in the blood and urine of the male proved little biologically, given that it was now acknowledged that the presence of follicular hormone was not exclusively linked to the presence of the female gonad. Despite having explained such a presence on the basis of Frank's theories earlier, 'Es éste hoy por hoy un problema turbador del que todavía no pueden emitirse afirmaciones concretas' (This is nowadays a troubling problem on the subject of which no concrete comments can be made). Whatever the origin of the female hormone, however, it could be affirmed that 'en el hombre con caracteres morfológicos intersexuales, lo cierto es que la foliculina se encuentra con más frecuencia en sangre u orina que en el hombre sexualmente normal' (in the man with intersexual morphological characteristics, what is certain is that ovarian follicular hormone is found in the blood and urine more frequently than in the sexually normal man).[191]

In the thought of Gregorio Marañón there is a constant tension between the ultimate mobility of sexual characters, their origins and the need to establish classificatory systems to determine precisely the nature of 'hermaphroditism' discussed, within a framework that acknowledges the porous nature of the sexes but also the desirability of acute sexual differentiation. However radical Marañón might appear in terms of his theory of intersexuality, sexual mixtures and continuums between the sexes, for the system to be able to operate it was predicated on the possibility of identifying 'male' and 'female' characters, sexes, hormones and histologies. We have seen how the cracks in this enterprise were numerous and, even by the admission of Marañón himself, there were many questions still to be answered.

It was perhaps these doubts that engendered a number of responses to Marañón's theory of intersexuality. In addition to the literary jibes that we referred to above, some criticisms were serious and well argued. F. Oliver Brachfeld, for example, accused Marañón of possessing 'nineteenth-century views' on the development of sexuality.[192] The pseudonymous Dr Baloardo criticized Marañón's theory of intersexuality in the form of a long poem, of which the following is merely an extract: 'Así pues, basta ya. Cese el empeño / de explicar con razones de paloma / los vicios de Sodoma / y hacer huertos floridos con ensueño / y ramas secas de podrido leño' (And so, enough! Let the attempt to explain with dove-like reason the vices of Sodom to make the gardens of dreamy flowers and bundles of rotten wood cease).[193] On the other hand, there were numerous adulations of Marañón's work and contribution to Spanish science.[194] This constituted Marañón's legacy – as open to question as the systems of classification by which he strove to identify 'intersexuals'.

Notes

1 Huerta (1929: 9) wrote 'Marañón es nuestro Darwin. Un Darwin *muy siglo XX*' (Marañón is our Darwin. A *very twentieth-century* Darwin).
2 Jiménez de Asúa (1934: 333).
3 The original article from the *Boletín del Instituto de Medicina Patológica* is reproduced in Marañón (1968c).
4 On the hormonal treatment of amenorrhoea see Sengoopta (2006: 159–63).
5 Marañón (1968c: 1044).
6 Ibid.
7 Ibid.
8 Dreger (1998: 10).
9 See Anon. (1931) where Dr Cardenal reports on a case of 'pseudo-hermaphroditism'. Marañón remarks (1968c: 465) that cases such as the one brought by his colleague are of great interest but 'no deben ser llamados de seudohermafroditismo, sino de verdadero hermafroditismo, o, mejor, como estados de intersexualidad que pueden variar desde el hermafroditismo al invertido sexual' (should not be called [cases of] pseudo-hermaphroditism but of true hermaphroditism, or, better still, as states of intersexuality which can range from hermaphroditism to the sexual invert).

10 Dreger (1998: 156) remarks that David Berry Hart, a Fellow of the Royal College of Physicians, called in 1914 for the elimination of hermaphroditism from the medical lexicon. Even 'pseudo-hermaphroditism' should be abandoned as a term since 'we cannot have a pseudo form of a non-existent condition'.
11 See the discussion in Cleminson and Vázquez García (2007: 29–94).
12 Marc (1817).
13 Ibid., pp. 89–90.
14 Ibid., p. 96.
15 Ibid., p. 117.
16 Perineal-scrotal hypospadias and its importance in cases of pseudo-hermaphroditism was discussed by Pozzi (n.d.: 586–92).
17 Ibid., pp. 583–92.
18 Cacho Viu (1997: 58–62).
19 Navarro Cánovas (1904) reported on a case discussed by Dr Katzenstein in the *Medizinische Wochenschrift*. This was a very rare instance of epispadias of the glans penis, only observed two or three times. This was the fourth case and perhaps the only one to be cured by an operation, it was noted. Note the increasing rarity of all conditions related in some way to the field of hermaphroditism.
20 Del Valle (1910) discussed a case from the Berlin Medical Society. Dr P. Marcuse reported to the Medical Society in November 1909 the case of a 6-year-old girl who had been entered in the Civil Register as such but whose sex was now in doubt. Very early pubertal development was detected but with no breast development, a 'kind of erectile penis' was present in the middle of the scrotal sac; the urethra emerged from the base of the penis. The child had no female characteristics whatsoever. It was reported that, given the social importance of cases such as this, it was lamentable that the Civil Register did not allow for the inscription of hermaphrodites, presumably as a temporary category to be 'resolved' later. In this particular case, the parents were advised that they should wait for a later sign such as menstruation, a 'marked sexual tendency' or the emission of semen. Other doctors present recorded different diagnoses. For example, Dr Gottschalk believed the child to be a girl.
21 Anon. (1906). Once more, in the Berlin Medical Society, Dr M. Katzenstein discussed identification and classification of cases of cryptorchidism as well as their treatment. No mention was made of hermaphroditism.
22 See, respectively, Anon. (1907, 1910a), both reporting from the Paris Society of Surgery on how to surgically intervene in such cases; Anon. (1909) recorded a 'pronounced' case in a subject who was reaffirmed in the Vienna Medical Society as a male because of the secondary sexual characteristics; in the same medical society Anon. (1910b) discussed a case of 'Hipospadias peno-escrotal'.
23 Codina Castellví (1907). The report was extracted from the *Presse Médicale*.
24 Nevado Requena (1906).
25 Ibid., p. 373.

26 Ibid.
27 Ibid., p. 374.
28 Ibid. Other cases of ectopic testicles and surgical techniques to deal with them were present in the *RMCP*. See, for example, Lozano y Ponce de León (1908).
29 Dreger (1998: 68–70). As we will see, some of Marañón's ideas were not that far from nineteenth-century embryology.
30 Nevado Requena (1906: 374).
31 Ibid.
32 Ibid., p. 375.
33 Irueste (1908). The medical authority was Dr Angulo at the Spanish Academy of Medico-Surgery on 6 March 1908.
34 Ibid., p. 395. Of continuing importance in cases of dubious sex was the confirmation or restoration of the ability to procreate, one of the main medico-legal issues attended to in the nineteenth century (see Ch. 1).
35 Sarabia (1908). Morales published his account originally in the *Revista Médica de Sevilla* and, given the quotation marks within which the text appeared, it can be assumed to be full and complete.
36 Ibid., p. 446, emphasis in original.
37 French doctors were not keen to alter the sex of the individual, thus condoning same-sex arrangements. English doctors seem to have been more favourably disposed to this. See Dreger (1998: 120–1, 125–6).
38 Sarabia (1908: 446).
39 Ibid., p. 447.
40 Ibid.
41 In France doctors accused midwives of the same ignorance. See Dreger (1998: 52).
42 Sarabia (1908: 448). Morales did not seem to be cognizant with what his colleagues had published on surgical techniques previously. See, in addition, Anon. (1910c). This was a report on the III Spanish Congress on Surgery (Madrid, 9–13 May 1910). Other cases are seen in Dr Wotan's report (1908) on Dr Royo.
43 Camacho Alejandre (1913: 265).
44 Ibid.
45 Ibid.
46 Ibid.
47 Ibid., pp. 265–6.
48 Ibid., p. 266. The quotations in the rest of the paragraph come from this page.
49 Ibid.
50 Ibid.
51 Ibid.
52 See, for example, the account of an ovary transplant in the United States in an article in the scientific news section by Rodríguez Morini (1900) and Serrallach (1908).
53 Camacho Alejandre (1913: 268–9). Dr Katzenstein's address to the Berlin Medical Society on 15 Nov. 1905 was specifically referred to,

summarized in the *Deutsche Medicinal* (sic) *Zeitung*, where an embryological account was given of the anomaly.
54 See, for example, Codina Castellví (1909) reporting from the Paris Surgical Society; Del Valle (1909) on a case in Vienna; Anon. (1910d) from Berlin, where, despite the title of the piece, a case of male pseudo-hermaphroditism was reported.
55 Del Castillo Ruiz (1910: 222–3).
56 See Dreger (1998: 86, 93). Laparotomy was first used in France for diagnosing sex in the early 1890s but did not become commonly used until the second decade of the twentieth century.
57 Ibid., p. 145.
58 Ibid., p. 247, n. 16, from where this whole paragraph is drawn. See also the critique in Magnan and Pozzi (1911: 233–4). As the title of the case study implies, attention was paid to the 'sexual orientation' of the individual concerned and of those in general. Asexual, homosexual or inverted and heterosexual hermaphrodites were all discussed in this piece (pp. 234–9). In Spain this element was much less at the forefront of analyses.
59 Dreger (1998: 146).
60 See the case notes written up by Peset Cervera (1879). Peset recounted the story of 'Consuelo S.' whose parents wished to alter the entry in the Civil Register to change the entry from female to male. The case was described as an error of sex since the 'girl' suffered from complete hypospadias and in whom the testicles were detected. It was advised that any rectification should take place once the 'time of the passions' emerged and then the sex that was most convenient to the girl should be adopted.
61 López Peláez (1916), in the review section of foreign medical journals, reported on E. Wilhelm's discussion on the civil state of the hermaphrodite. The original discussion arose in the pages of the *Münch. Med. Woch.*, 9 (1916). All quotations in this paragraph are from this source.
62 Anon. (1920b). Klebs's categories can be seen in Dreger (1998: 145).
63 Anonymous review (Anon. 1923) of Knud Sand's work.
64 Pascual (1920: 181). All quotations in this section are from this study. This case appears in full in Cleminson and Medina Doménech (2004: 89–90).
65 We have already commented on the use of laparotomy. See Dreger (1998: 86, 93). Reference to the use of the microscope was made in Ch. 3.
66 Cardenal (1924). The case is recorded in full in Cleminson and Medina Doménech (2004: 90–1). It was also reported on more fully by Dr Maestre (1924) as 'Un caso de hermafrodismo'. The difference in terminology employed hardly by now requires much comment. Throughout the account given here some aspects from the *Siglo Médico* report will be commented on. Dr León Cardenal, professor of surgery at Madrid, worked closely with Marañón and in 1926 made testicular transplants in accordance with Voronoff's techniques on an impotent man and a eunuchoid adolescent. See Glick (1976: 294).

67 If more details could have been garnered in this sense a project more in touch with the excellent Oosterhuis (2000) could have been undertaken.
68 Cardenal (1924: 439). In the translation here we have used 'he' although in the original Spanish the pronoun is not gender specific.
69 In Maestre (1924) it was noted that there existed three kinds of true hermaphroditism: bilateral, unilateral and alternate, following Klebs's scheme without naming him. Dr Cardenal asked how it was possible for the person to possess secondary female characteristics and have male genitalia. This was explained by the 'tendencies' and 'psychological and somatic characteristics' of the patient being those of a woman. It was also noted that the hormones in men and women may well be identical. There was, however, another explanation to allow for the presence of traits of both sexes. The present case, Cardenal wrote, could have been a case of true hermaphroditism in her early years. In animals such as pigs the gland of one sex degenerates leaving just the other sex. This could also occur in humans. The secondary female characteristics may have arisen out of the early presence of the female gland, Cardenal wrote. Such an explanation, however, fixes the glandular action according to sex, something Cardenal had already questioned above. Another explanation would be that the female and male gland somehow became mixed, thus giving rise to both sexual types. This set of explanations concurs with those of Marañón (1972: 568).
70 He was also reported having said (in Maestre, 1924: 183) that in the testicle the internal and external secretions can be distinguished histologically. Masculine characteristics were attributed to the presence and action of the Leydig cells and it would be important to see if these cells existed in this case. Dr Cardenal replied that under the microscope, such cells could indeed be found. The male interstitial gland existed but the existence of female interstitial tissue may have led to the emergence of female characteristics. Compare once more with Marañón (1972: 575).
71 This point was elaborated upon by Maestre (1924).
72 Vara López and Sánchez-Lucas (1925).
73 Here we draw on an anonymous (Anon., 1925) review of Salvador Pascual, 'Los hermafroditas', *Archivos de Endocrinología y Nutrición*, 1 (1925).
74 Anon. (1925: 564).
75 Of the much-reduced attention paid to this topic we can cite Anon. (1919) in the review section of foreign journals in *El Siglo Médico*. Dr Adlercreutz in the *Nord. Med. Ark.*, Stockholm, presented the case of a girl of 13 who was then 'made into a boy'. The penis was freed by a number of operations. No reference was made to hermaphroditism. Also, see Morales (1923). The article fitted the kind of discussion on hypospadias earlier in the century. Dr Morales related the case of 7-year-old Agustina who had such a degree of hypospadias that parents were shocked to learn that she was in fact a boy (p. 549). Dr

Morales discussed what he understood to be a collective atavism in the village of Los Lagares (Malaga province). There were no schools in this rural province and the peasants lived a 'practically savage life' (p. 550). In order to avoid military service the *lagareños* baptized their sons as daughters, which in turn created a difficult legal situation if they wished to marry.

76 Vandebosch (2006: 81).
77 Aresti (2001: 88, 92–3). Aresti notes that female commentators such as Emilia Pardo Bazán and Carmen de Burgos Seguí saw the First World War as a watershed in changing awareness. On gender relations in workers' organizations, see Nash (1995: 7–42).
78 Aresti Esteban (2007: 176) quotes Jiménez de Asúa, *Al servicio de la nueva generación* (1930) on the threat to home and maternity that many men believed the 'new woman' posed.
79 Ibid., p. 177 quoting Saldaña, *Siete ensayos sobre sociología sexual* (Madrid, 1929).
80 Charnon-Deutsch (1996: 78).
81 Aresti (2001: 60–1) citing Roberto Novoa Santos, *La indigencia espiritual del sexo femenino (Las pruebas anatómicas, fisiológicas y psicológicas de la pobreza mental de la mujer. Su explicación biológica)* (Valencia, 1908).
82 Fernández (1996: 84).
83 Litvak (1979: 150).
84 Benita Asas Manterota, 'De las nuevas culturas', *Mundo Femenino*, 81 (1932), p. 4, cited in Aresti (2001: 240).
85 Vandebosch (2006: 11–14).
86 See Ortega y Gasset (1983).
87 Rafael Abellán, *Historia crítica del pensamiento español*, vol. 5/3, *La crisis contemporánea. De la Gran Guerra a la guerra civil española (1914–1939)* (Madrid: Espasa-Calpe, 1991), p. 63, cited in Vandebosch (2006: 12).
88 See the prologue by Marañón (1929: 13) to Hernández-Catá's *El Ángel de Sodoma*.
89 Glick (2005).
90 In the *Evolución de la sexualidad y los estados intersexuales*, Marañón (1972: 607) remarks favourably that Spain was one of the first nations to remove punishment of homosexual acts from its statute books.
91 Marañón (1967a). The speech was given in Seville and was reproduced in the *Siglo Médico* in 1920.
92 Marañón (1967a: 10).
93 Ibid., pp. 12–14. Cf. Geddes and Thomson (1889) discussed in Weeks (1989: 146–7).
94 Marañón (1967a: 13). Marañón repeated many of his ideas in different contexts. This notion can be seen, for example, in his 'Sexo, trabajo y deporte' (Marañón, 1967b: 105–6), a conference speech originally given in 1925.
95 Marañón (1967a: 27). Here Marañón followed the Italian edn of Otto Weininger's *Sex and Character* (Milan, 1912) to justify his remarks. The Spanish edn, *Sexo y Carácter*, was published in 1902.

96 Marañón (1967a: 28). This phenomenon was described by Marañón as 'partial heterosexualism', that is, the combination of characters of the two sexes (nothing to do with 'heterosexuality' as such).
97 Marañón (1968a).
98 Ibid., p. 171.
99 Ibid., p. 172.
100 Aresti (2001: 221, n. 48). This was not the only literary filter through which Marañón was distilled. Wright (2004: 737) introduces the 'bawdy vaudeville farce' of *La plasmatoria*, which played with notions of intersexuality and anxieties over masculinity invested in Don Juan. See Pedro Muñoz Seca and Pedro Pérez Fernández, *La plasmatoria: farsa cómica en tres actos* (Madrid, 1936).
101 Aresti (2001: 95).
102 We have already noted that Dr Cardenal utlized Voronoff's techniques. See the discussion in José Madinaveitia, 'Las operaciones de injertos gandulares en la Facultad de Medicina', *El Sol* (1 March, 1926), cited in Glick (1976: 294).
103 Glick (1976). For a good general introduction to the subject see Hall and Glick (1976). Here the authors argue that endocrinology is not a specifically defined discipline but rather a field, as it drew on a wide variety of resources to argue its claims.
104 See the discussion of Monlau ('el útero hace que la mujer sea lo que es' (the uterus makes the woman what she is) in his *Higiene del matrimonio* (1853), cited in Fernández (1996: 100).
105 Oudshoorn (1994: 8).
106 Ibid., p. 19.
107 Ibid., pp. 15–16, 19–20.
108 Ibid., p. 21. Oudshoorn (1990: 176) discussing developments in the late 1930s, remarks that even though the idea of sex chromosomes being agents of sex determination had been proposed in 1906, techniques for detecting them were not yet available: 'In this context it can be understood that the expectations were high that sex hormones would provide scientists with a tool to determine the sex of hermaphrodites and to explain the "feminine" character of homosexual men.'
109 Oudshoorn (1994: 22).
110 Quoted in Hall (1976: 273).
111 On this process see Borell (1976; 1985: 3).
112 Quoted in Nelly Oudshoorn (1990: 166).
113 Borell (1976: 266).
114 Borell (1985: 5).
115 See, respectively, Rodríguez Morini (1900); Serrallach (1908).
116 Glick (1976: 288).
117 Ibid., p. 294.
118 See the discussion in ibid., p. 290.
119 Marañón (1966a). This material forms Marañón's acceptance speech into the Royal National Academy of Medicine in March 1922.
120 Ibid., p. 15.

121 See Glick (1976: 290). Gley's talks were published as *Quatre leçons sur les sécrétions internes* (Paris, 1920). Gley's criticism also turned on Marañón for his 'uncritical' use of organic extracts in the clinic. Marañón defended himself from Gley by alleging the latter's 'lack of perfect comprehension of the (original) Spanish text' of Marañón's *Las glándulas de secreción interna y las enfermedades de la nutrición* (2nd and 3rd edns), mentioned in Marañón (1966a: 15–16, 26, n. 1). Medvei (1982: 393–5) discusses this crisis in Spain and follows Glick's analysis closely.
122 Marañón (1966a: 89). In Spain, developments did not necessarily follow these steps. Glick (1976: 287–8), following Nicholas Mullins, has argued that a different chronology was established whereby a new paradigm was taken up by a few individuals, next a communication network was established, then a cluster of interested scientists and finally the institutionalization of a specialty took place. In Spain, the last step occurred when Marañón became head of the Institute of Pathology, discussed above. See Mullins (1972).
123 Marañón (1966a: 29–60).
124 Ibid., p. 41. Marañón wrote that even Gley had acknowledged that the notion of genital internal secretions was the best established of all (ibid.).
125 The notion of primary and secondary characters was invented by John Hunter in 1870 according to Aresti (2001: 121).
126 Marañón (1966a: 42).
127 Ibid.
128 Ibid., p. 43.
129 Ibid.
130 Ibid., p. 74.
131 Ibid., p. 88. Marañón's optimism with respect to the power of the internal secretions to determine all behaviour was muted in later essays. In his prologue, 'La endocrinología y la ciencia penal', to the criminologist Q. Saldaña's book *Nueva criminología* (Marañón, 1936: 10), he admitted that the effects of the internal secretions were rarely decisive.
132 Oudshoorn (1990: 169).
133 Ibid., pp. 169–70.
134 Ibid., p. 170. On the basis of these findings, some clinicians argued that those human subjects where the 'wrong' hormone was found were latent hermaphrodites (Oudshoorn, 1994: 27) and many saw the presence of the female hormone in the male body as an agent of disease and disorder, in particular 'psychosexual disorders' such as homosexuality (p. 32). In the discussion of case number 9271 in his Institute of Medical Pathology (Marañón, 1932a) different doctors referred to endocrinological issues (Marañón and Planelles) and to psychic factors (Garma). For Marañón the lack of male hormone and the fact that the female hormone occurred in the urine of the subject confirmed 'de modo sensacional el diagnóstico de homosexualidad' (sensationally the diagnosis of homosexuality). What it

confirmed, however, must have in reality remained unclear since the male hormone was absent in many cases of sexually normal men and the female hormone occurred in the urine of normal men.
135 Oudshoorn (1990: 171).
136 Oudshoorn (1994: 23, 32). In the mid- to late 1930s more complicated systems emerged whereby it was thought that were at least two different types of female hormones; others divided them into male, female and bisexual hormones in both sexes. The international conferences on the standardization of hormones (1932, 1935) attempted to resolve these complications (Oudshoorn, 1994: 35, 46).
137 Marañón (1972: 499–710).
138 Marañón (1967c). Papers of similar content were also given in various places in Spain throughout 1928 and 1930 (ibid., p. 185).
139 Ibid., p. 156.
140 Ibid., p. 157.
141 Ibid., p. 159.
142 From now on Marañón refers regularly to Goldschmidt's legacy. The latter's first major paper on the subject was 'Intersexuality and the endocrine aspect of sex', *Endocrinology* (Philadelphia), 1 (1917), 433–56, referred to in Marañón (1972: 509, n. 1).
143 Goldschmidt (1917: 437, n.).
144 Marañón (1972: 509, n. 1).
145 Marañón (1967c: 175), emphasis in original.
146 Ibid. The biological and evolutionary thought that hermaphroditism was an atavistic or primitive trait is clear here. See Ch. 1 for a discussion of this point.
147 Marañón (1967c: 175).
148 Ibid.
149 Ibid., p. 176. Marañón also elaborated on these rather more social methods in Marañón (1926a, 1926b).
150 For example, in Marañón (1968a). Marañón wrote: 'No es que un sexo se desarrolle y el otro desparezca. Lo que ocurre es que uno se hipertrofia y el otro se atrofia' (It is not that one sex develops and the other disappears. What happens is that one hypertrophies and the other atrophies) (p. 165).
151 Marañón wrote of the 'superación' (overcoming) of the feminine by the masculine (ibid., pp. 180–1). What happened, then, is that '*Los dos sexos no se oponen ... sino que, sencillamente, se suceden*' (*The two sexes are not opposed to one another ... rather, quite simply, one follows on from the other*) (p. 182, original emphasis).
152 Ibid., p. 169. Sexual differences were seen as deeply rooted. In those women who displayed a 'feminidad muy pura' (very pure femininity), their libido would be 'indiferenciada y confusa' (undifferentiated and confused). The male, on the other hand, would sport a 'libido enardecida [que] se dirige a la hembra con violencia radical y directa' (enflamed libido [which] is directed towards the female with great direct violence) (p. 175).
153 Ibid., p. 168. These findings were summarized at a later date. See Marañón (1968b: 199–200).

154 Marañón (1972: 560).
155 Ibid., p. 566, original emphasis.
156 Ibid., p. 567, original emphasis.
157 He acknowledged that this was so in the cases discussed by Pascual in the *Archivos de Endocrinología y Nutrición*, 3 (1925), p. 13, and by himself in *Trabajos del Servicio de Patología Médica del Hospital General de Madrid*, 4 (1927–8).
158 Marañón (1972: 568).
159 Oudshoorn (1990: 172–3) notes that Robert T. Frank who wrote *The Female Sex Hormone* (Springfield, IL and Baltimore, MD, 1929) had found the female sex hormones in 'normal' 'healthy' males but that these originated from food.
160 Marañón points to Soler y Dopff and Forcada Gelabert, *Revista Médica de Barcelona*, 10 (1928), p. 217, as an example of this.
161 Marañón (1972: 572). He noted also that 'masculine' libido might also intensify in older age. Marañón made few comments on the perceived sexual orientation of his hermaphrodite subjects, somewhat surprising given his extensive studies on homosexuality and 'sexual inversion'.
162 Ibid.
163 Ibid., p. 575.
164 Ibid.
165 Oudshoorn (1990: 180) writes that Carl Moore and Dorothy Price at the University of Chicago extended the conceptualization of sex from the gonads to the brain. In 1932 they postulated the existence of a feed-back system between the gonads and the hypophysis, suggesting that 'the inhibiting effects of female sex hormones on male sexual characteristics could not be understood in terms of a direct antagonistic effect on the male gonads, but rather in terms of a depressing effect on the hypophysis, thus diminishing the production of male sex hormones by the gonads'.
166 Marañón (1972: 576). Marañón declared that his new system had been accepted by Dr Recasens, in the *Revista Española de Obstetricia y Ginecología* (Valladolid), 14 (1929), p. 32.
167 Marañón (1972: 576).
168 Ibid., p. 578.
169 Ibid., p. 584, original emphasis.
170 Orcoryen (1929: 500).
171 Torre Blanco (1929: 529).
172 Ibid.
173 Ibid., p. 530. Other examples of the reception of ideas on endocrinology in the same journal are: Gil y Gil (1929) and García Triviño (1929). On a different note, 'Progynon', the female sex hormone 'according to Prof. Steinach, Vienna' was available for purchase from an outlet in Barcelona. This advert appeared in many issues of the *AMCE*.
174 Marañón (1928) reporting on the clinical session in Marañón's Service of Medical Pathology in Madrid of 6 Oct. 1928. This would be

the case that was also discussed in the in-house bulletin *Anales del Servicio de Patología Médica del Hospital General de Madrid*, 4 (1928–9), pp. 3–4 (Wright, 2004: 728).
175 Oudshoorn (1990: 174).
176 Usandizaga and Sánchez Lucas (1931).
177 Ibid., p. 373.
178 Ibid., p. 375.
179 Marañón (1972: 566–74) in the section on hermaphroditism did not refer to Hoepke.
180 Usandizaga and Sánchez Lucas (1931: 376).
181 Ibid.
182 Ibid., p. 377.
183 Ibid., p. 378. This line of reasoning evidently followed Marañón. See Marañón (1972: 575).
184 Anon. (1931).
185 Ibid., p. 465.
186 Ibid. Dr A. Gimeno Cabañas was evidently of a more traditional stance. He related the case from 1881 of a male who considered himself as such and who tried to escape military service passing himself off as a woman. He had always had a 'masculine' sexual appetite. This was a case of 'pretended', 'intentional' or 'supposed' hermaphroditism (the original Spanish can be read in different ways). See Gimeno Cabañas (1931).
187 See the discussion in Marañón (1972: 568).
188 Marañón (1931a) discussed on 28 Nov. 1931 at the National Academy of Medicine.
189 Ibid., pp. 587–8.
190 Marañón (1931a: 588).
191 Ibid.
192 Brachfeld (1932), and, Marañón's response (1932b). Brachfeld's view of Marañón as outdated comes from his *Polémica contra Marañón* (Barcelona: Sob. de López Robert y Campanía, 1933), p. 42, cited in Wright (2004: 729). See also Criado y Aguilar (1932a, 1932b).
193 Dottore Baloardo, 'Cronicón académico: la intersexualidad y el regodeo cientificoide', *El Siglo Médico* (28 March 1931), p. 346, cited in Aresti (2001: 221–2). Other examples of criticism appear in this section of Aresti's book (pp. 221–6). This must have been going on for some time as criticism of 'Dr Baloardo' and his 'más pintorescas' (most folkloric) opinions by thirty-four scientists who signed a collective letter had appeared on 23 Feb. 1929 (Various authors, 1929). Marañón, Novoa Santos and Cardenal were among the signatories.
194 Among the many see the two reviews of his work by Dr Francisco Haro García. The first was of Marañón's *Tres ensayos sobre la vida sexual* (Haro, 1926) and the second was of Marañón's *Los estados intersexuales en la especie humana* (Haro, 1929).

Chapter 5
From True Sex to Sex as Simulacrum

Over the period 1930–1970 Spanish medical thought would undergo a number of crucial changes with respect to notions of sexual identity. These changes amounted to the installation of a new 'truth regime' with regards to sexuality during this period. From the period of enlightened despotism through to the liberal revolutions of the nineteenth century, the medico-legal profession had consolidated its idea of sexual dimorphism and had attempted to fix clear criteria for the classification of cases of doubtful sex or apparent hermaphroditism.

Going beyond the markers of male or female sex provided by custom and behaviour, forensic doctors searched tirelessly for physical proofs of maleness or femaleness. From the end of the nineteenth century, in the light of the work of Klebs and Pozzi, they appeared to locate the truth of sex in the histological examination of the gonads of the individual. In this way, an 'age of gonads' began,[1] which in Spain arrived late and had to compete with a new style of thought emerging around 1913: reference to the hormones as the new marker of sexual identity. Sex was still thought to be based on a material reality but not on any one 'thing', the gonads, but on a process that at the same time biological and chemical. Sexual difference was produced by the complex dynamic of the internal secretions.

The most successful version of this hormonal explanation of 'true sex' in Spain was that provided by Marañón and his followers from 1920 onwards. However, as we saw in Chapter 1, the work undertaken by the Dutch team headed by Ernst Laqueur and the publications of the German gynaecologist Bernhard Zondek at the end of the 1920s and beginning of the 1930s cast doubt on the sex-specificity of sex hormones.[2]

The evidence against the sex-specificity of the hormones accrued over the years that followed. Hormones no longer offered a stable account upon which to judge true sex. The concept of true sex was displaced by that of socially 'convenient sex'; the materialist vision was replaced by one that was more pragmatic by the end of the 1950s.³ It was accepted that it was possible for there to be discordance between the scientific determination of sex (as based on the chromosomes) and the effective designation of sex. The latter was thought to derive more from psycho-social learning processes than from a person's biological personhood. This kind of psycho-social reasoning, pioneered by the American psychologist John Money,⁴ was taken up in Spain by the team directed by the gynaecologist Botella Llusiá.

Despite what one might think at first sight, the end of true sex as a biological reality did not mean the end of the dualistic understanding of the two sexes. The imperative of 'one body, one sex' was maintained as impregnable. What was admitted was the disassociation between the biological sex and the psycho-social sex, an understanding that, amongst other factors, allowed for the emergence of the category 'transsexual'. It was the set of psycho-social factors that were deemed most important in the designation of sexual identity. But at the same time, it was argued that the balanced mental health of any individual required strong, early and exclusive socialization in the ways of one sex and one sex alone. In this way, any transgression of the conventional roles associated with each sex, including 'sexual orientation', was considered to mark a failure in the process of sexual development and the cause of psychopathological problems later in life.⁵

In accordance with these ideas, it was imperative that the sexual appearance of the individual be modelled as early as possible on the characteristics of the sex that had been assigned from birth. In the case of doubtful individuals, this meant that the old obsession that looked for signs of the true sex of the individual in the body now turned to a different kind of embodiment of sex via surgical means. These would assist in the creation or reassertion of a socially acceptable appearance according to sex, either male or female. The regime of true sex was replaced by the regime of sex as appearance or simulacrum. In this chapter, we trace how this transformation took place in Spain. The analysis will discuss the various stages of this process together with the changing gender representations and surgical developments during the period from the Second Republic through to the late period of the Franco regime.

Intersexualities

In Spain in the early 1930s the science of sexual identity was dominated by the thought of Gregorio Marañón. But despite the predominance of his theories of 'intersexuality' they were not the only explanation available. As we saw in Chapter 4, he had many critics, both among those who defended a strict biological dichotomy between the sexes against the notion of intersexuality[6] and those who criticized what they thought to be Marañón's endocrinological reductionism and his lack of attention to the psychological dimension of sexual identity.[7] Marañón's thought itself also evolved over time and his initial enthusiasm for an all-embracing endocrinological explanation shifted towards a more restricted role for the power of internal secretions.[8]

Be that as it may, the thought of Marañón on sexuality transcended the scientific community to exert during the early Republic a huge social impact with a large degree of uptake in scientific and social elites in the big cities. His thought, in addition to its scientific dimension, offered a liberal and humanistic understanding of the nature of sex and sexuality and the new expressions of both that shook the traditional borders between the sexes[9] – the presence of the 'new woman' in the 1920s in women's circles and in the artistic and journalistic circles of Madrid and Barcelona,[10] feminist movements, which had gained strength in the same decade and which were very active during the Republic,[11] and the increased visibility of a subculture of 'sexual inverts' which in turn had its own voice and increased presence in the cities, novels and vanguard poetry.[12]

Marañón adopted an original strategy in order to respond to these new doubts around strict sexual identities. Instead of reaffirming the dualism of the sexes by restricting, as did those who still favoured a gonadal explanation for sex, the existence of 'true' hermaphrodites to extreme cases, Marañón widened their presence to cover a whole range of figures that Goldschmidt had classified under the category of 'intersexuality'.[13] Cryptorchid and hypospadiac males, bearded women, hermaphrodites with ovo-testes, virile women and homosexuals and a huge range of more or less exotic intermediates were grouped together in the vast spectrum of intersexuality.

Instead of reaffirming sexual dimorphism as a simple category, however, Marañón's multiplication of sexual types had the objective of protecting the differences between the sexes by relegating any transgressions (feminism, homosexuality, new women) not to the field of perversions, sin or vice,[14] but to the anomalous and varied field of intersexuality; such a move, paradoxically, had the effect of reinstating sexual dimorphism as, if not fully attainable, at least desirable. Such a move also allowed Marañón to effect a partial break from the old misogynist discourse on the biological inferiority of women, to defend the decriminalization of homosexuality and to consider maternity as the culmination of womanhood.

At the same time, the distance between 'normal' and intersexual persons was reduced. The process of sexual differentiation experienced regressions and interruptions at crucial moments such as puberty in males and in the climacteric in females. Marañón and his researchers also believed that sexual differentiation was an unfinished process and that no individual could claim to have achieved full differentiation.[15] However advanced this hormonal process, likened to the action of a driver with the gonads that guided the 'motor' of the sexual cycle,[16] may have been there would always remain some elements of the other sex.

At the beginning of the 1930s a group of researchers coalesced around Marañón (such as Planelles, López Escoriaza, Pardo, García Orcoyen and Gómez Acebo) and assiduously reported upon their activities with respect to intersexuality undertaken in Marañón's own institute and in the Spanish Medical-Surgical Academy. The activities of these research groups are published in, for example, the *Archivos de Medicina, Cirugía y Especialidades*, and bear witness to the attempt to create an interdisciplinary debate on the question. Not only physiologists participate in these debates but others from a broad range of fields, with gynaecologists such as Vital Aza, and renowned psychiatrists such as Rafael Garma making their presence felt.[17]

In these sessions, it was accepted as a principle that the key to sexual differentiation resided in the internal secretions, which were based, in turn, on the differences between male and female hormones. However, the biological substrate to the question, which had been so important in the 1920s in Marañón's thought, having relied on the work of scientists such as Lipschütz, Bauer, Steinach and Goldschmidt, amongst others,[18] faded in significance in the light of the chemical explanations articulated by the Amsterdam school directed by Ernst Laqueur and by the researchers of gynaecologist Bernhard Zondek at the University of Berlin.[19]

On an experimental front, the controversy over the effects of the use of gonadal insertions in castrated individuals gave way to a debate on the presence, composition and therapeutic use of hormones. In this respect, histological and surgical developments were replaced by bio-chemical and pharmacological interpretations and practices.[20]

On the biological level, Marañón's circle accepted the difference between sex determination, as governed by the chromosomes, and the process of sexual differentiation, which depended on the action of the internal secretions of the gonads. Despite this, from the early 1930s, Marañón began to accept a role for other endocrinal glands in the shaping of sexual characteristics. This broadened analysis permitted an escape from the dilemmas of the previous decade, which had set geneticists (who argued that sex was established only by the action of the chromosomes) against physiologists (who understood sex to be affected at least in part by the environment of growth and the glandular system of the embryo's development).[21] The original sex of the individual did not depend, therefore, on the gonad alone; it was established by what Marañón called the 'danza complicada' (complicated dance) of the chromosomes.[22] The function of the hormones was confined to protecting the chromosomal determination of sex ('como los pastores al rebaño' (as shepherds protect their flocks)),[23] guiding it in its development through the complex original 'bisexuality' of the embryo.[24] The various intersexual states would result from disorders in the action of the glands.

This notion was further complicated by what Marañón called a 'problema turbador' (disturbing problem),[25] however, as pointed out in the previous chapter. The hormones were found not to be sex-specific but were found at least in some proportion in both sexes, according to the researches of Zondek and Laqueur. The principles that constituted the science of endocrinology in its early period from 1910 were no longer tenable. Sexual hormones and, therefore, sex itself, were not rooted in any anatomical or functional base, unlike gonads. This interpretation was substituted by a relational approach that 'delocalized' the process of sexual differentiation, situating it diffusely in the whole endocrinal system.

By the 1930s several pieces of evidence for this idea had accumulated. Glands such as the hypophysis,[26] the thyroid[27] and suprarenal glands,[28] whose hormones acted directly on the activity of the gonads and on the process of masculinization and feminization, were held to play a vital role. The action of these glands brought about fundamental revisions of the supposed all-important role of the gonads and their secretions and placed at the forefront what began to be termed 'functional correlation'[29] or 'synergy' and which in later years would be termed 'feedback' and its role in the process of producing sexual differentiation.

In this way, the simple, linear causality of the gonads was replaced by a complex, circular action of the whole of the glandular apparatus. Such a theory also explained the relationship between the intersexual states and conditions such as dysfunctions linked to the metabolism,[30] pregnancy[31] or lactation.[32]

This breakdown of the relationship between locality and the processes of sexual differentiation also brought about a 'desexualization' of the whole process. If the sexual hormones, in terms of male and female hormones, were no longer thought to condition the sexual evolution of the individual, did it still make sense to speak of sex-specific hormones? This separation from previous paradigms was encouraged by the discovery of progesterone. Instead of the dualistic notion of male and female development being led, respectively, by the male androgens and the female follicular hormone, Zondek's work showed that sexual differentiation did not respond to a double process but a triple one: the action of the hypophysis on the sexual cycle of the female, which had been investigated more fully than the male because of its relationship with maternity and reproduction, was in turn dependent on the action of two types of hormone. First, Prolan A, that aided in the maturation of the follicles, acting on ovulation and, secondly, Prolan B, that stimulated luteinization, related to pregnancy.[33]

To some degree, of course, the reduction in the importance granted to the gonads vindicated Marañón's schema of broad intersexuality in contrast to the exceptionality of the true hermaphrodite as evoked by the followers of the gonadal theories (Klebs, Pozzi and Neugebauer). But what was in reality most confusing and inconvenient for the notion of 'sexual' gonads was the presence of female hormones produced by the testicles and male hormones produced by the ovaries. The existence of these 'heterosexual hormones', as they were called, broke down the division between maleness and femaleness in the realm of the internal secretions.

Marañón and his followers vindicated this cross-over between the sexes and the hormones as proof of their theories of intersexuality.[34]

The presence of large quantities of female hormones in the urine of males and of male hormones in females did indeed seem to prove their claims. On the basis of this presence, the group began to try to locate large quantities of follicular hormone in the urine of eunuchs,[35] gynaecomasts,[36] and homosexuals,[37] as well as the activity of masculinizing hormones in virile women.[38] Marañón did acknowledge that attention should be paid to the work of Robert Frank published in 1929 whereby the massive presence of follicular hormone in men was adduced to metabolic alterations as a result of the ingestion of certain foods. But this admission in itself did not cast doubt on the theory of intersexuality; on the contrary. Individuals with these alterations, such as the 'Eskimos', who ate a lot of fish did indeed show frequent manifestations of intersexuality.[39]

What did contradict Marañón's theories, however, was the presence of huge amounts of follicular hormone in 'normal' males, that is, those with no homosexual tendencies and with no evident intersexuality.[40] These facts, asserted by several researchers since Zondek in an important article published in 1934, thoroughly questioned Marañón's readings. Zondek argued that the follicular hormone present in the urine of males came about as a result of its conversion from a male hormone into a female hormone.[41] If the gonads of both sexes produced both male and female hormones and if both could be converted into the other, it could not be sustained, as Marañón had done to date, that between both forms of internal secretion there existed fundamental antagonism and a primordial incommensurability. From this time on, the idea of the antagonism between the sexes was steadily replaced with the concept of *cooperation* of both in the same individual.[42] In the final instance, Marañón's idea of sex dualism with respect to the hormones was not sustainable.

Furthermore, many of the certainties of Marañón's theory were placed in doubt in the clinic and in practical diagnoses with respect to the determination of sex. As we have seen, such an examination could be performed as a result of doubt in the case of the recently born or infants, or as an accidental consequence of a routine surgical or medical intervention, such as for hernias.

In an example of how medical theory does not always coincide with the practice of medicine (similar examples can be seen in the history of psychiatry), doctors in cases of doubtful sex usually relied on gonadal criteria during this period and on the taxonomy established by Pozzi, who had distinguished between pseudo-hermaphrodites of two types, the least frequent gynandryne and the androgyne. This occurred in four case studies reported upon in the medical press during the Second Republic.[43]

The first is a case of a new-born baby who was baptized as a girl and who was examined in 1931 by Dr Muñoyerro Pretel, a doctor attached to the Inclusa (child hospital and orphanage) in Madrid, an association which would have given him access to a large number of cases. Muñoyerro would rise to the position of Head of the Provincial Service of Child Hygiene.[44] The case is classified by Muñoyerro, following Pozzi, as an androgynous pseudo-hermaphrodite boy, as the presence of testicles was discovered in the scrotum. On a theoretical level, however, the doctor concurred with the thought of Marañón, considering that hermaphrodites and pseudo-hermaphrodites should be classed as variations of degree on the scale of intersexuality.[45]

The second case was published by Mas Collellmir in 1932. He too describes to a new-born baby, on this occasion in Barcelona province, which was registered as a boy. Shortly after birth, the parents noticed some genital abnormalities and sought medical advice. On examining the baby, the medical doctor, whose doctoral thesis was dedicated to the subject of intersexuality, found a nodule that seemed to correspond to the shape and qualities of a testicle. Citing Neugebauer on the frequency of such cases, Mas Collellmir classified the baby as a male pseudo-hermaphrodite, although he reserved final judgement for a later date as he had not been able to make a histological examination of the testicle. Once again, as in the case discussed by Muñoyerro, this Barcelona doctor, drawing on Steinach and Marañón, argued that the concept of 'pseudo-hermaphroditism' was inadequate, since 'no existe un pseudo-hermafroditismo o hermafroditismo falso en el verdadero sentido de la palabra, sino que todos los estados intersexuales pueden ser esbozados o completos, pero son siempre verdaderos' (pseudo-hermaphroditism and false hermaphroditism do not exist in the true sense of the word. Instead, all the intersexual states can be either partial or complete but they are always true).[46]

Our third case comes from 1934 and is reported by the Catalan doctors, Pons Tortella and Gállego Berenguer. A 23-year-old single woman from Teruel was admitted to the Provincial Hospital for an operation on her appendix. During the intervention, an ectopic testicle was found and, in addition, her male appearance, hirsuteness, lack of menstrual flow and masculine voice all made the doctors suspect a true male sex lying beneath. The woman was diagnosed an 'androgynous pseudo-hermaphrodite', that is, a male who appeared at first sight to be a female, although no legal mechanisms or measures were suggested to rectify her civil identity.[47]

Our last example, published in 1936, was reported twice by Muñoyerro Pretel.[48] This time, Muñoyerro was presented with the case of a boy of seven days, who was in the Provincial Institute for Puericulture suffering from ictericia and skin scaling. An initial observation of the external genitalia recorded the presence of a rudimentary penis and of scrotal sacs. Although no testicles were recorded, the initial diagnosis followed Pozzi and Neugebauer and the boy was classed as one of 'pseudohermafroditismo masculino o androginoide' (male or androgynous pseudo-hermaphroditism), although the author preferred Marañón's more simple 'niño pseudohermafrodita' (pseudo-hermaphrodite boy). After a bout of influenza which developed into pneumonia, the child perished. On performing the autopsy, fully developed internal female organs were discovered. Such a discovery obliged the doctors to reverse their diagnosis – instead they had a case of a 'pseudohermafrodita ginandroide' (gynandroid pseudo-hermaphrodite) or a 'niña pseudohermafrodita' (pseudo-hermaphrodite girl). As in the 1931 case Muñoyerro entered into a long digression which explained

> *según las modernas orientaciones de Marañón, Bauer, Lipschütz, se trata más bien de verdaderos hermafroditas, es decir, que contienen las dos gónadas, pero con predominio de una, quedando la del sexo contrario relegada a segundo término, como adormecida, latente, estando siempre alerta y en acecha para aprovechar momentos oportunos y hacer acto de presencia.*

according to the modern teachings of Marañón, Bauer and Lipschütz, rather they are cases of true hermaphrodites, that is, they contain the two sets of gonads but one is predominant. The other sex is relegated to second place, as if asleep, latent, but always alert and lying in waiting in order to take advantage of opportune moments to make its presence felt.[49]

Evidently, there was a huge gap between the practical diagnosis and the theoretical framework invoked to explain hermaphroditism. Many medico-legal treatises did in fact admit this difference,[50] which was justified for two reasons. First, any medical examination was obliged to resolve quickly any doubtful cases of sex and assign a stated sex for the individual concerned. Secondly, any scientific conclusions could often only be reached, in an era when exploratory surgery or laparotomy was just beginning to be employed,[51] once the patient had died or, when alive, with his or her permission. This was especially the case in private consultations.[52]

Such a schism explains the coexistence of contradictory conceptual accounts of hermaphroditism, that is, those taxonomies associated with a set of strict gonadal criteria, which restricted the possibility of 'true hermaphroditism', and those that followed Marañón's theories of intersexuality, which multiplied the possibility of a hermaphrodite diagnosis. The first set of criteria was invoked to resolve cases on a practical level while the second was employed to explain hermaphroditism theoretically.

Hermaphrodites in early Francoism: revival and crisis of the gonadal criteria

It is now commonplace for historians of early Francoism to emphasize the totalitarian bio-political nature of the state that grew out of the Civil War.[53] This new state aimed to exterminate any remains of a liberal, lay, leftist tradition, understood as part of the 'anti-Spain' exemplified in the Second Republic.[54] This undertaking to create afresh was often expressed in a form of language that was biological and religious at the same time, couched in terms such as 'purification' and 'regeneration'.[55] The national body or organism had to be rid of 'germs' and 'bacteria' that had infected the nation during the republican period.

One of the most important fronts on which this purifying battle would be fought was that of scientific research and university life, held to have been contaminated by the 'plague' of free-thinking and laicism encouraged by the Institución Libre de Enseñanza.[56] Intellectual circles were subject to a process of 'purification' or purge (*depuración*), which gave rise to what one author has called an 'atroz desmoche' (appalling beheading),[57] a decimation of Spanish scientific research that set the country back by decades.

Medical research and teaching were not exempt from this process.[58] The names Emilio Mira, Juan Negrín, Severo Ochoa, Rodríguez Lafora and Marañón himself,[59] are just the tip of a huge iceberg of people obliged to seek temporary or permanent exile.

Libraries, research centres, departments and institutes were dismantled in this frenzy. The reading of accusations before the Tribunal of Political Responsibilities (February 1939) by erstwhile colleagues seeking the professorial chairs of those accused is an example of the worst kind of political vengeance afforded by the regime.

This desire to 'purify' Spain and effect its rebirth was accompanied by an exhaustive process of 're-Catholicization' of all aspects of public and private life from schools through to national holidays and dress codes. Episodes such as the rise of the new woman, the advances of the feminist movement, the visibility of homosexuality, were all seen as products of the degeneration that had been promoted by the years of republican rule.[60] Rigid distinctions between the sexes were the order of the day. This division of roles became the hallmark of the new Christian family, whose only legitimate expression in terms of sexuality was reproduction.[61]

On this subject there was ample agreement between the religious and new civil powers of the Franco state. Procreation as a function of marriage was seen as inherent to Catholic teaching as well as integral to the objectives of the new state. In similar fashion to the pronatalist policies of Italian fascism and National Socialist Germany, the power of the nation was predicated upon the biological quantity and quality of its population. Demography, social medicine and the racial policy[62] of the new state incorporated the concept of 'Geburten-soll', taken from the thought of Third Reich social scientists in order to designate the proportion of children that each family should generate in order to guarantee the biological and political future of the 'national community'.[63]

It is in this light that the opposition to 'degenerate' practices in sexuality are to be understood and the extirpation of abortion, neo-Malthusianism and contraception was undertaken. A whole raft of legislation (laws to protect the birth rate, against abortion considered as a crime against the state, against contraception, adultery and infanticide, against divorce, in favour of family subsidies, prizes for especially fecund marriages and protection for large families)[64] was aimed at annihilating the old bio-political order of the 'anti-Spain' to achieve the restoration of Imperial Spain. This pro-natalist push in the context of the post-war demographic collapse also aimed to combat high rates of infant mortality by means of the consolidation of fields such as maternology, puericulture and public hygiene. A whole range of measures were undertaken in order to change the health and morality of the people.[65]

Evidently, medical knowledge had a huge part to play in this new undertaking and it was harnessed to reinforce the particular perspectives of the Catholic Church and the regime. But medical knowledge of the hormones, for example, was not simply rejected outright. At the same time as matrimony[66] and maternity[67] were exalted as duties established by biological and religious laws, and Marañón's ideas on intersexuality were criticized as undermining the essential sexual differences as set out in the Creation,[68] research into the therapeutic use of hormones was undertaken as a way of resolving sterility in married couples.[69] A whole range of problems, including impotence, premature ejaculation, hypospadias, cryptorchidism, endocrine hypo- and hyperactivity, were analysed against the backdrop of sterility.

Gynaecologists and obstetricians, amongst others, concentrated their efforts in the struggle against sterility as a patriotic endeavour.[70] These efforts came together in the field of maternology. In 1942 twenty maternity institutes existed in the whole of Spain, five of which were based in Madrid. Between 1940 and 1942, 180,000 women attended prenatal sessions at these centres.[71]

Such developments enable us to understand the new directions taken with respect to the question of hermaphroditism in the new regime. During the 1940s medical figures attempt to distance themselves from Marañón's thought on intersexuality. Sex differences, from a bio-medical perspective, are reinforced. In terms of the hormones, we see a return to notions of sex-specificity.[72]

In addition, and in connection with the crusade against sterility, we find a lively interest in endocrinological research into the influence of the internal secretions on pathologies of reproduction.[73] If during the 1930s the emphasis had been placed on research into the 'female' hormones, during the 1940s there is a rise in coverage in Spain of attempts to obtain and synthesize 'male' hormones by medics such as Butenandt, Ruzicka and Laqueur.[74] The failure of surgery based on gonadal inserts is recorded[75] and there is growing interest in the therapeutic use of synthetic hormones above all, but not exclusively, in the field of sterility.[76]

Another peculiarity of the medical publications of this period is the tendency towards producing once more a stigmatized representation of those individuals who were sexually ambiguous, including eunuchs,[77] and women undergoing the menopause.[78] Within this context, the therapeutic use of androgens is rejected in women, despite some advantages offered in the treatment of some tumours, as it was felt they caused 'una innecesaria, incómoda y patológica virilización' (an unnecessary, uncomfortable and pathological virilization) of women,[79] and that their use could be a cause of problems in pregnancy: 'especialmente está contraindicada en el embarazo, por llegar a producirse el aborto o la intersexualidad del feto, si aquél no llega a interrumpirse' (it is especially not advised during pregnancy as the abortion or intersexuality of the foetus may arise, if the pregnancy itself is not interrupted).[80] The maintenance of strict divisions between the sexes may well have been more important than the survival of the mother concerned. Despite this concern to fix strict boundaries between the sexes, in accordance with the new regime's ideology, doctors could not avoid having to accept eventually the non-specific nature of the sex hormones and that hormones of all types could produce masculinizing or feminizing effects.[81]

Given that the hormones did not seem to provide any reliable basis from which to apportion sex differences, it is not strange that many doctors reverted to a renovated version of the old histological criteria: true sex would be designated by the true gonad. Such a shift is apparent in some studies on hermaphrodites and pseudo-hermaphrodites in the 1940s in Spain. In these medical reports, a new discursive strategy appears that was not present in the 1930s. On the one hand, doctors affirm the extremely delicate nature of their cases and request that such cases be shrouded in silence and left to the appropriate specialist.[82] On the other hand, there is a return to a very 'nineteenth-century' rhetoric; these studies emphasize the impossibility of 'true hermaphrodites' in the human species.[83] These reports are positioned at the opposite extreme from the thought of Marañón, which tended to expand the range of intermediate figures.

The first of the reports that betray this return to older postulates is that of the Catalan doctors Simarro Puig, Otero Sánchez and Lluch Caralps in 1942. Before discussing the case in question, the authors enter into a long theoretical exposition on the taxonomy of hermaphroditism. The framework advanced by Pozzi is accepted, whereby distinction is made between gynandryne and androgyne pseudo-hermaphrodites, and his thought is praised as bestowing 'practical' advantages because of the 'simplicity' of his schema.[84]

The principal theoretical referent in this case, however, is not Pozzi's work. Instead, it is the monograph published by the French surgeon Louis Ombrédanne, *Les Hermaphrodites et la Chirurgie* (Paris:

Masson, 1939).[85] This author, who also subscribed to Pozzi's classifications, understood that in human beings there was no such thing as true hermaphroditism since there had never been found any case of an individual capable of fertilizing and being fertilized at the same time. This rather reductionist set of criteria allows the authors to reconcile the claims of science with those of the Catholic faith, which only allows for two distinct sexes. In addition, such an understanding coincides completely with the pro-natalist stance of the regime – what determines sex is the ability to procreate. The gonadal criterion is, therefore, re-established as a means of determining the true sex of the individual but in a different way. The key is no longer seen to reside in the anatomical or histological structure of the organs but rather in its function.[86]

The case in question focuses on an individual of 18 years who is registered as a female. She comes to the clinic because 'desde hace algunos meses se halla sujeto a frecuentes humillaciones a consecuencia de su aspecto varonil' (for some months she has been subject of frequent humiliation on account of her manly appearance).[87] She declares that she is attracted to women and suspects that her true sex is male, something which 'desea que sea así' (she desires is the case).[88] The Catalan doctors, following Ombrédanne, highlight the importance of the psychological aspects of the case and the secondary characteristics. There is no mention of any external female sexual organs and the bodily image is defined as completely masculine. The only female element is the nature of the pubic hair. The presence of supposedly functioning testicles is detected, given the declarations of the individual who attests to ejaculations on rubbing a small penis or 'over-sized' clitoris.

The subject in question also spoke to the doctors about how, some years previously, it had been suggested in a Barcelona surgery that she undergo an 'intervención plástica con objeto de "formar una vagina artificial"' (plastic intervention with the objective of 'forming an artificial vagina'),[89] on the belief that she was in fact a woman. Finally, the diagnosis comes that the doctors are seeing a case of 'perineal hypospadias', which in earlier years had 'simulated' pseudo-hermaphroditism. The doctors revert back to Pozzi and they state that they accept provisionally the latter's classification of androgyne and gynandryne pseudo-hermaphrodites. But in this case, 'ni siquiera puede hablarse de androginoide, ya que los caracteres sexuales secundarios eran propios como menos que totalmente del sexo real del individuo' (one cannot even talk of a androgyne as all the secondary characters were either mainly or totally those of the real sex of the individual).[90] Evidently, the authors' position that true hermaphrodites in the human species are not possible is confirmed. Such a piece of scientific truth would nonetheless support 'los preceptos religiosos' (religious precepts).[91] The masculine nature of the individual was confirmed thanks to the 'examen del líquido o substancia emitida en las pérdidas anotadas' (examination of the liquid or substance emitted in the losses referred to).[92] What the doctors do not explain, however, is how such a sample was obtained, something delicate for the period concerned.[93] What remains clear is that the *function* of the gonads, in this case, the production of sperm and hence the possibility of fertilization, was the key that allowed classification as a male.

For the regime and the doctors concerned, ambiguous individuals threatened the rigid dichotomy between the two sexes. The authors of this report emphasize, in particular, the risks entailed by such ambiguous traits in the school environment as illustrated by Ombrédanne.[94] In the Francoist education system, which was opposed completely to coeducation, intermediate figures had no place. For this reason, once the true male sex of the patient was determined, the doctors advise, apart from correcting the Civil Register, a 'terapéutica quirúrgica ortopédica dirigida a corregir su malformación urogenital' (orthopaedic surgical therapeutic intervention with the object of correcting their urogenital malformation), to be accompanied by subsequent testosterone treatment.[95]

The second case we analyse from the 1940s was published in 1943 by Dr Pedro Piulachs, a specialist in surgical pathology at the University of Barcelona. The case focused on a girl of 7 years who was taken in as a result of a suspected hernia. During the ensuing operation a testicle the size of a chickpea was found. Laparotomy allowed Dr Piulachs to confirm the lack of internal female organs despite her external female appearance. A second testicle was found and was positioned alongside the first in the periostium of the pubis. The testimony of the mother pointed to no obvious masculine behaviour on the part of the girl. She was assigned the male sex, such was communicated to the family, and she began to be dressed in accordance with her new male identity. The boy, whose photograph is reproduced in the report, was then transferred to a playgroup organized by the Auxilio Social, the Francoist social relief programme. Three years later he was subjected to a psychological and psychometric examination. This test showed a certain educational backwardness and an overly affectionate character but nothing significant in terms of sexual identity as at this stage the 'atracción sexual de los púberes' (sexual attraction of adolescents)[96] was not yet present.

In this case, the theoretical part of the report, which was much shorter than that of the first case, followed the physical description. Piulachs drew upon the typologies configured by Klebs and Pozzi. In contrast to the case discussed by Simarro Puig et al., there is no strict negation of the possibility of true hermaphroditism.[97] Once more, the author relies on Ombrédanne: the key to the 'true sex' of the individual does not reside in the macro- or microscopic morphology of the gonads. The primary sexual characters (the external genitalia), the secondary characters (the internal genitalia) and the tertiary characters (bodily traits, behaviour and character) are also relevant. But

> *por encima de la valoración de las formas cree [Ombrédanne] que debe darse prepondenrancia [sic] a las funciones, aunque no siempre pueden éstas ser claramente determinadas. En realidad, como hecho funcional de valor decisivo sólo habría el embarazo para la mujer y la eyaculación de espermatozoides vivos, para el hombre.*

over and above the importance of the form he [Ombrédanne] believes that function should take precedence, although the function cannot always be clearly determined. In reality, as a fundamental element of decisive value only pregnancy in women and the ejaculation of live sperm in the man count.[98]

Once more, in the context of profound pro-natalist concerns, the functional criteria are foregrounded as the most important element. The hormonal aspect is not even mentioned as if it provided no clear basis for the determination of sex. Femininity was equivalent to the capacity to give birth and care for children. Birth and reproduction were for women and were biological and patriotic obligations; a numerous progeny, for men, was deemed evidence of a strong male spirit.[99]

The third case we look at from the 1940s points to a new theoretical departure.[100] It was published in 1944 by doctors E. Roda, A. de la Peña and E. de la Peña, all urologists at the Madrid University Medical Clinic, under the prestigious pathologist Jiménez Díaz. One of the peculiarities of the clinical example examined, a case of 'masculine pseudo-hermaphroditism', is that the article begins with an extensive biographical account of the individual, giving her, although indirectly, a certain degree of voice. The case is of a 30-year-old woman, born in Nogales (Lugo), who was sent to the urology service from her home provincial hospital, because of the presence of 'un cordón espermático y un testículo hipoplásico' (a spermatic cord and a hypoplasic testicle).[101] The patient arrived dressed as a woman and remarked that, according to her mother, the matron had considered her to be of doubtful sex on birth. Despite this, she was baptized and registered as a girl.

A description follows that is based on the patient's own account. She refers to her masculine inclinations in terms of the toys she played with and in respect of her 'libido'. Here, she recounted her inclinations towards the female sex. The repression of these impulses led to their 'la satisfacción solitaria' (solitary satisfaction), that is, masturbation,[102] although she did indicate that she had on occasion tried to cohabit with women. This had been difficult because of her genitalia. This 'double life' led her doctors to think of their patient as subject to bouts of make-believe, an interpretation often espoused in the face of hermaphrodites who recounted tales of deception, fraud and hiding of real identities.

After this biographical section, the medical examination is discussed. The subject appears in the *Revista Clínica Española* with her face fully uncovered. The examination is exhaustive, including details of her voice, manners, bodily traits, and the appearance of her external and internal genitalia, something facilitated by laparotomy. A variety of techniques was used including X-rays, metabolic analyses and radioscopy. However, in the absence of the available equipment, no hormonal analysis took place.

In the conclusions where the case is described as one of 'male pseudo-hermaphroditism with hypospadias' the authors present an extensive account of current thought in hermaphrodite science. They believe that the classic taxonomies of Klebs and Pozzi do not conform to recent developments in gynaecology and genetics, although their 'utilidad en clínica' (usefulness in the clinic) is accepted.[103] In fact, the diagnosis relies on these old criteria, despite the doctors' critique of them, but not without some changes. They prefer the use of the term 'intersexuality' over that of 'hermaphrodite' and 'pseudo-hermaphrodite' and refer to Goldschmidt and to Marañón to sustain their stance, while reserving much praise for the latter's volume *La evolución de la sexualidad*.[104]

One novelty that appears in the piece is the importance granted the chromosomal mechanism in sex determination. The authors, although they did not possess the techniques required to perform the genetic diagnosis of sex, refer in detail to such procedures. Together with the predominant role accorded to the sex hormones in the embryo stage, they consider that the chromosomal structure is the decisive factor in sexual differentiation.[105] They point out, nevertheless, that the variety of intersexual states is potentially very broad and complex given the varied effect that endocrinological factors could have upon sexual differentiation. In turn, the varied effects of the glands and the lack of specificity of the hormones could well entail 'influencias que parecen paradójicas' (effects that appear to be paradoxical).[106] By this the doctors meant the feminizing effects of testicular extracts and the masculinizing consequences of the presence of oestrogens. Once more, the lack of sexual specificity of the sex hormones seems to hint at the undermining of the natural division into two distinct sexes.

One of the reasons behind our categorization of this text as transitional is its reference to the gonadal criteria as a function rather than a morphological state. This understanding is invoked in order to reject the possibility of hermaphrodites in the human species.[107] Furthermore, the reference to true biological sex is maintained, which is thought to reflect the true identity of the person. However, despite these rather older assertions, this text is one of the first to insist at the same time on the importance of the chromosomes; it is also receptive to Marañón's ideas on intersexuality and, above all, it emphasizes the importance of the psychological and social process of learning in the construction of sexual identities.

The authors reiterate that it would be 'absurd' to assign sex purely on the basis of the presence of male or female gonads.[108] Those individuals with somatic characteristics, such as the external genitalia, and psychic elements of the female sex, for example, although they may possess testicles should not be converted into males. What should take place is the 'extirpación de los testículos con administración simultánea de estrógenos' (extirpation of the testicles with the simultaneous administering of oestrogens).[109] The key here is the 'aspecto físico y psíquico del individuo' (physical and psychic aspect of the individual), not the gonads they possess. In the case in question, the doctors suggest that the sex is changed in the civil and canonical registers and that dress, appearance and occupation should be altered in accordance with the 'new' sex. In addition, 'está justificada la terapéutica encaminada a favorecer, en lo posible, la máxima diferenciación sexual, y por tanto viril, del individuo' (therapeutic techniques directed towards promoting, as far as is possible, maximum sexual differentiation of the individual in a virile sense are justified).[110] Not only should surgical intervention discover the real sex of the person; it should also help to reinscribe it.

Despite these recommendations, the urologists discover that their patient had continued to live as a 'mujer en su indumentaria, trabajo, etc.' (woman in her dress, occupation, etc.).[111] This attitude, which may today be understood as an attempt from a transgender perspective to refuse sexual pigeon-holing, is discussed at length in the article in the context of the huge psychological cost that renouncing one's learned sex and adopting one's 'verdadero estado' (true state) would entail.[112] Others, however, the authors remind their readers, 'encuentran la felicidad en el cambio de sexo' (find happiness in their change of sex).[113]

This emphasis on the relevance of social learning and the individual's own psychological welfare, as well as the comments on the importance of genetics will be confirmed in subsequent pieces in the Spanish medical press.[114] In the meantime, medico-legal manuals continued to refer to the usefulness of typologies such as that of Pozzi and – in an environment in which sexual ambiguity was contemplated as an unacceptable threat – they continued to associate hermaphroditism with perversion and criminal behaviour.[115]

Between chromosomal sex and apparent sex: sex determination in authoritarian Francoism

Throughout the 1950s the tendency towards accepting a distancing between true biological sex and socially assigned sex, something already emerging in the previous decade, was consolidated. The work of the influential gynaecologist José Botella Llusiá,[116] who had already been significant in the campaign against sterility[117] and in research on the endocrinological causes of intersexuality,[118] was also important in the 1950s in forging new ways of examining cases of dubious sexual identity. The materials that Botella Llusiá published from 1953 onwards on the subject of various clinical cases of hermaphroditism and pseudo-hermaphroditism introduced a new type of language that was distanced from the biologism of the immediate post-war period. In these texts Botella conceded that biological sex, even as a chromosomal category, did not tell the whole story about sexual identity in ambiguous cases. Other factors needed to be considered such as the 'dignity', 'satisfaction' and psychic 'welfare' of the person concerned in order to avoid their being cut adrift socially and personally. For this reason, Botella always advised that the civil sex, that is, the sex of appearance that had been learned socially and inscribed in any register, should be the one to prevail in any assignation of sex.

Before analysing the work of Botella it is necessary to reflect briefly on the social and political circumstances of 1950s Spain. As is well known, the end of the Second World War meant extreme international isolation for the Franco state, given its co-operation with the Axis powers. The United Nations confirmed this pariah status in 1946 by recommending no diplomatic relations with Spain. From the mid-1940s the regime tried to free itself from this cul-de-sac by reaffirming its anti-Communist credentials in the context of the emerging Cold War and by playing down any existing fascist rhetoric. Internally, this meant the sidelining of pro-Nazi Falange elements by Catholics and others in a rejigging of the relative power of the components of the 'families' of the regime. Many Falangists from the end of the 1940s and beginning of the 1950s altered their stance in order to fit in with this changed reality.[119] Several rather cosmetic constitutional changes resulted, including the Fuero de los Españoles of 1945 in an attempt to make Spain presentable to the international community.[120]

North American support would also be decisive for the opening up of a second reform process from 1956 onwards after the first great university protests and the major strike wave of 1951. A new cabinet of 1957 was largely populated by technocratic members of Opus Dei, bringing in a degree of economic liberalization under the Stabilization Plan of 1959. Both the old Falange-inspired bellicosity and traditional Catholicism were substituted by new forms.

The need to incorporate increased numbers of workers into a developing tertiary sector served to soften opposition to working women and spurred on the reform of discriminatory laws against women.[121] In addition, the take-off of a consumer economy and a growing middle class in the 1960s, together with the massive receipts gained from tourism, occasioned large changes in customs and lifestyles and challenged the ascetic life of post-war values and the rigid dichotomies between the sexes.

This backdrop of changes introduced from the mid-1950s onwards allows us to reassess the novelty of the thought of Botella with respect to cases of 'confused' sexual identity. In an early article of 1953 Botella examined the protocols and possibilities of surgical intervention with respect to cases of ambiguous sex.[122] Botella classified intersexuals in accordance with an understanding in tune with the work of Havelock Ellis. As such, he distinguished between the primary characteristics (the gonads), secondary characteristics (the genitalia) and tertiary elements (apparent bodily traits, from voice to appearance). In this way, he rejected any attempt to find an organic locus for sexual identity. This was understood not to lie in the gonads or in the hormones; the latter were chemically similar across the sexes. Sex was understood to inhabit the whole organism because the process of sex determination 'es eminentemente genético' (is eminently genetic).[123] Botella conceded that intersexuality could occur without there being a strict chromosomal basis, as in cases of hormonal imbalance, but the most extreme example of an intersexual state, true hermaphroditism, always depended on a genetic foundation.

Botella attempted to uncouple the popular association between intersexuality and homosexuality. He believed that, amongst intersexuals, homosexuality was exceptional. This allowed him to introduce a further explicatory reflection on the subject of the 'sexual instinct'. The sexual instinct was always understood to arise from 'factores educacionales y sociales' (educational and social factors),[124] and was not much affected by the hormones or other biological factors. This thesis was defended at the time of a certain degree of panic around the association between homosexuality and political subversion.[125] Homosexuals were understood to form part of a secret lodge which was dedicated to undermining the foundations of western Christian society. From this perspective, the biological conception of homosexuality was lost, a conception that refused to blame the homosexual for his condition. Instead, in Spain the rejection of what was essentially Marañón's idea on homosexuality was made explicit in this move.[126] The notion of blame under Spanish National Catholicism was reaffirmed explicitly. The 'dangerousness' that these individuals entailed – soon to be recognized in law[127] – did not free them of personal responsibility. Homosexuals were classed as individuals who gave themselves over to 'vice' and who, of their own free will, in accordance with traditional Catholic thought, committed crimes against God and the nation.

The same kind of 'Christian humanism' with regard to the free choice of the individual, rejecting in this way any kind of biologism or racial essentialism so common in the 1940s, is what drives Botella's thought on sex-change surgery. In those cases where the gonadal sex, the external genital morphology and the sexual instinct all coincided 'correctly', the surgeon should not intervene to change sex. The surgeon should merely try to correct any imperfections that caused confusion for the patient's identity. Such intervention did, however, include, for example, 'corrigiendo su hipospadias o creando una vagina, según el sexo' (correcting hypospadias or creating a vagina, according to sex).[128] But these were corrections rather than sex changes. In the case of true hermaphrodites any intervention should follow the social sex that the individual presented and any signs of the opposite sex should be eliminated surgically.

Where did pseudo-hermaphrodites fit into this picture? Their position was, from a deontological perspective, more complex. In these cases, Botella argued that the social sex that these individuals had been assigned should be maintained: 'manteniendo al sujeto en el sexo civil y ocultando, sobre todo cuando los interesados lo ignoran, la naturaleza verdadera del sexo' (maintaining the subject in their civil sex and hiding, above all when those in question are unaware of it, the true nature of their sex).[129] This hiding of the true sex of the individual patient posed a difficult moral question for the doctor. The Madrid gynaecologist tackled the issue by referring to three female cases that had come under his auspices. The three came to his clinic complaining of hernias. As was common in such cases, the women in fact possessed testicles and were ignorant of their state. Their sexual desires were directed towards men. The option taken was to castrate them and to keep silent on their true biological sex. In order to justify such acts of medical paternalism, Botella referred to the psychological welfare of those affected, and emphasized, above all, the social consequences that the revelation of the truth would entail.¹³⁰ Once more, the social panic around homosexuality rears its head:

> *si declarásemos varón a una de estas mujeres, con aspecto físico completo de mujer, sería a todas luces un ser desgraciado, objeto de burlas y que no podría casarse. Sus instintos femeninos, debidos al factor fenotípico, que, siendo creídas mujeres, eran los normales y correctos, persistirían en su nuevo estado, convirtiéndolas en homosexuales y creando una continua fuente de perversiones y conflictos.*

If we were to declare one of these women to be a male, with their completely womanly physical appearance they would be the object of humiliation and would not be able to marry. Their female instincts as a result of their phenotype, which, given the fact that they were believed to be women, were normal and correct, would persist in their new state and they would become homosexuals, thus creating a continuous source of perversion and conflict.¹³¹

The hiding of the patient's 'true' sexual identity was preferable to the encouragement or consolidation of any homosexual tendencies.¹³²

The text displays an acute degree of anxiety over the possibility of altering the sex assigned to the individual socially. Rectification of sex is not always rejected as a theoretical possibility,¹³³ but this course is viewed with extreme caution given the social consequences and ensuing scandal that such a change of sex would entail. The division between the sexes, fixed *ab initio* by the Creator, did not respond to any secure basis in biology.¹³⁴ Ethical and juridical considerations should be aware of the 'mystery' of the sexes that could not be resolved by mere science alone;¹³⁵ hence the call for a form of pragmatism guided by social convenience. The truth of the dichotomous nature of the two sexes remained as unassailable; what confirmed this was not biology but the demands of civilization.

Ambiguity and sex changes were not rejected on the basis of degenerative traits or racial stigmas. What we see instead is a curious mix of language that refers to intersexuals' 'inadaptación social' (lack of social adaptation)¹³⁶ with explanations that draw on notions of political subversion.¹³⁷ One case that exemplifies this kind of thinking is that of Teresa Pla Messeguer (Villabona, Castellón, 1917), the so-called 'hermaphrodite maquis'. According to official legend, Pla Messeguer, 'La Pastora', formed part of the anti-Franco guerrilla movement in the Sierra del Maestrazgo in the 1950s and was responsible for the murder of numerous persons and for a raft of robberies.[138]

Teresa was born with ambiguous genitalia, although with dominant masculine characteristics. Her father, acting on the advice given by a friend, decided to register her as a girl in order to avoid any eventual problems that could have been encountered during military service. In the 1930s Teresa became a shepherd in the mountains of the Sierra del Maestrazgo and dressed as a woman. However, she always felt she was a man and maintained sexual relations with some young women in the area. In her village she was renowned as a tomboy or 'marimacho'. After an incident in 1949 with the Civil Guard she became a guerrilla and took the name Florencio and the pseudonym 'Durruti'. The Civil Guard encountered her caring for her sheep on the mountain slopes and, having heard of her ambiguous traits, forced her to undress and go down on all fours so they could examine her. This event, together with the setting fire to a barn where two maquis had sought refuge, convinced her to enrol with the guerrillas and she fled to the mountains. She fought over a period of twenty months and participated in some actions but without causing injury. Subsequently, she took refuge in Andorra where she earned a living by means of smuggling and shepherding until caught by the Civil Guard in 1960.

With the reputation as a blood-thirsty guerrilla that the press had created – some authors described her as a 'lesbian woman with violent instincts' – she was to fall victim to the authorities' need for a scapegoat for a number of unresolved crimes. Teresa was found guilty and condemned to death but her sentence was commuted to thirty years in prison. First of all, she was delivered to the women's prison and was forced, despite her beard, to don a tight-fitting dress. After being examined by military medico-legal doctors she was transferred to the men's prison. She left prison in 1977 and obtained a change in identity, being registered as a male.

The undertakings of La Pastora were strictly contemporaneous with the studies by Botella on hermaphroditism. While he did not discuss, as far as we know, La Pastora herself, his 1953 studies discussed the three cases of male pseudo-hermaphroditism outlined above.[139] In 1956 Botella returned to the question and this time analysed two cases of 'true hermaphroditism'.[140]

Botella made known these cases in a speech at the Royal National Academy of Medicine in an analysis that pondered the abundance of hermaphrodites in the Mediterranean basin. Of a total of sixty true hermaphrodites registered, four had been from Spain, 'en donde, sin embargo, y por razones sociales, el pudor a describir tales anomalías es extraordinariamente grande' (where, however, for social reasons, the shame entailed by recounting these anomalies is extraordinarily acute).[141] Botella presents the different types of intersexuality in accordance with the three types of sexual characters as elaborated by Havelock Ellis. True hermaphroditism, therefore, corresponds to the hybrid nature of the gonads; pseudo-hermaphroditism occurs where there is an ambiguity in the conformation of the relevant distinct sexual organs and virile and feminine states are related to alterations in the tertiary characteristics.

The first case described a single male, 19 years old, an agricultural labourer, who comes to the clinic complaining of a measure of discomfort in his genitalia. He also notes huge growth of his breasts to the extent that they now appeared to be female breasts. His libido is masculine, that is, directed towards women, and he wishes to remain as a male. On rectal examination, one of the methods favoured when laparotomy was not possible, it is detected that an ovary is present and that the subject possesses one testicle.

The second case is more complex. A young woman who had been sent to the clinic on the orders of a magistrate and who had been accused of the rape of a minor was discussed next. She was accused of deflowering a young servant girl of the same age in the house where the accused also worked. The girl had a feminine appearance, regular menstruation, and possessed a clitoris of some 5cm in length. She had a narrow vagina that impeded sexual relations with men. An ovo-testis was extirpated, thus apparently confirming the diagnosis of true hermaphroditism. Except for her sexual orientation (directed towards the same sex) she is, in the eyes of the doctor, completely feminine. For this reason, the psychiatrist that examines her believes that 'sería un error convertirla en varón por medios quirúrgicos' (it would be a mistake to convert her into a male by surgical means).[142]

Botella, after conducting a hormonal analysis in both cases, concludes that the causes of hermaphroditism must be genetic. Taking into account other similar clinical cases, he considers that an analysis of a buccal sample would reveal cells whose chromosomes were in all probability female.[143] Drawing on established medical literature, Botella argues that these subjects may well have suffered some kind of chromosomal mutation from XX to XY during the embryonic state, thus resulting in the alteration of gonadal differentiation and true hermaphroditism.[144]

What is perhaps most striking when reading this text is the attention paid to the psychic state of the patients and the deontological implications of this kind of case. Both cases reaffirm the lack of connection between sexual preference and biology.[145] The first case, brought up as a male, had developed a sexual instinct towards men. The second case was attracted to members of her own sex. Botella believes that this sexual orientation resulted from the physical impossibility of 'normal' sexual relations given the shape of the vagina. Homosexuality, in this case, would not have emerged spontaneously but as a compensation for an impossible expression of heterosexuality.

When it came to suggesting surgical intervention, Botella, as in the cases of pseudo-hermaphroditism discussed in 1953, argued that it was imperative to adjust the patient to his or her 'civil sex'. If the doctor acted in order to reinforce the sexual appearance that the individual had learned and grown up with 'problemas de adaptación no se presentarán' (problems of adaptation will not arise).[146] This criterion which favoured civil sex was reinforced by the assertion that it was necessary to go with the sex that the individual believed he or she possessed. This would allow for 'la satisfacción del propio sujeto al poder desarrollar aquellas tendencias sexuales que en él ya existían' (the satisfaction of the subject himself as he would be able to develop the sexual tendencies that were already present),[147] and would avoid sudden changes of sex that could lead to suicide.[148] Botella's speech finished by recalling the three cases of pseudo-hermaphroditism discussed in 1953 and called once more for the concepts of 'dignity' and 'happiness' to be the cornerstone of all interventions.[149] The irony, however, of Botella's leaning towards the conception of sexuality that the individual had him- or herself is that he may have endorsed, in a manner of speaking, homosexuality as a legitimate sexual option.

These criteria upheld by Botella and which were in general subscribed to by international medical opinion on the subject would be applied in those cases published in the Spanish medical press at the beginning of the 1960s.[150] Two cases we now consider referred to a female case[151] of pseudo-hermaphroditism and a male case.[152] A third example focuses on a case of true hermaphroditism.[153]

The female pseudo-hermaphrodite was examined by a team of medical doctors from the Institute of Clinical and Medical Research (Instituto de Investigaciones Clínicas y Médicas) directed by Carlos Jiménez Díaz, one of the period's eminent figures. Here we have the case of a 33-year-old man who was married and was a labourer. Chromosomal analysis and laparotomy showed genetic and gonadal female sex. However, because of what appeared to be the presence of congenital androgenital syndrome, caused by suprarenal hyperplasia, the individual presented an external appearance that was completely masculinized. The libido was also male and the individual affirmed that he experienced orgasm with his wife. Following the protocol outlined by Botella, the doctors in charge decided to remove the female genital organs, which were the equivalent of those of a girl of 10, and to insert a simulacrum of male genitalia. After this successful intervention, the patient was placed on a course of hormonal therapy. The presentation of this case as a social problem, given the civil status of the individual, justified acting in a way that would leave 'esta criatura en completa ignorancia de su problema' (this creature in complete ignorance of her problem).[154]

The case of male pseudo-hermaphroditism examined by a team of doctors at the University Medical Clinic, also directed by Jiménez Díaz, appeared to present fewer complexities. The doctors examined a 45-year-old male from Palencia. His general morphology (breasts, distribution of fat, hair, etc.) was feminine but his libido, gonads and chromosomal sex were declared to be male. He was diagnosed as a case of 'hereditary male pseudo-hermaphroditism' caused by a pathological gene that resulted in a hormonal imbalance. It was decided to retain his sex as male.

The third case, studied by doctors at the medical clinic at the Faculty of Medicine in Barcelona, unusually corresponds to a case of true hermaphroditism. The authors of the report adjust this concept somewhat, stating that the true androgyne was a product of mythology and not reality; 'realmente nunca se ha producido en la especie humana' (it has never in truth arisen in the human species).[155]

They go on to discuss an individual aged 10 who was registered as a male. The analysis of the sex chromosomes revealed a female genetic sex. Laparotomy, however, revealed the presence of an ectopic testicle and of an ovary that was in reality an ovo-testis. This was a case of true hermaphroditism whose anomaly was attributed to 'una traslocación de genes durante la meiosis con la formación de un gameto en el que un cromosoma X llevaría genes masculinizantes procedentes del cromosoma Y' (a transference of genes during meiosis with the formation of a gamete in which an X chromosome would possess male genes coming from the Y chromosome).[156] The patient was diagnosed with Klinefelter's syndrome.

In their consideration of the psycho-social aspects of these cases, the Spanish doctors in question drew support from a study from 1955 published by John Money and his team at the Johns Hopkins University.[157] These studies argued that sexual identity depended upon the upbringing of the child. From the age of 18 months the child would begin to develop its own awareness of self and its own idea of the sex it belonged to. From this age onwards it would be increasingly difficult to modify psychologically the subject in order to infuse him or her with the sexual identity contrary to that with which they were brought up.[158]

The boy in question had no erotic desires but since he had been brought up as a boy it was this sexual identity that was thought should prevail and be fostered. The doctors who discussed his case were clear that, in their view, any therapeutic action or corrective surgery should not take precedence over or contradict 'las aspiraciones del individuo' (the aspirations of the individual).[159] As the Franco state's system of medical care shifted from a regime whereby the health of the individual was seen as a duty to the *patria* towards the logic of a welfare state, where health was understood as an individual demand for a particular service, it is not surprising to see an increased emphasis on the availability of corrective surgery in order to achieve the 'sexo deseado' (desired sex).[160] Doctors recognized that it was easier, technically speaking, to construct an artificial vagina and to extirpate a 'penis-like' clitoris than it was to make a scrotum and a penis that would be 'apto para la cópula' (adequate for coition).[161] But in this case, given the fact that the desired sex, which was also the one that was socially prevalent, was the male sex, the medics decided to extirpate all the gonads of either sex, prescribing androgens and placing in the scrotal area 'dos "testículos" de vitalio o de material plástico' (two 'testicles' of artificial material or plastic).[162]

The three cases mentioned here, in addition to illustrating the thought of Botella Llusiá throughout the 1950s and the decline of homophobia in the 1960s, are interesting for two reasons. First, Botella's thought shows how Spanish medical figures were as knowledgeable in technical and epistemological terms as their international peers on the question of hermaphroditism. But they were not culturally prepared for the arrival of the notion of 'transsexualism', which was becoming accepted in the United States.[163] Despite this, given that it was possible to accept the disassociation between biological sex and psycho-social sex and to concede greater importance to the latter, in reality Spanish doctors were just one step away from accepting in a biologically normal individual the presence of a psychological 'illness' such as transsexualism or 'gender dysphoria'.[164] If the desired sex was to be given predominance in intersexuals, why could not desired sex operate in the same way in individuals who were psychically damaged by not living in a body they experienced as their own? If nineteenth-century legal medicine had made the body the prison of the soul, as truth was seen to be located in the body, the new medicine of the sex change had made the soul a torturer of the body. The body became something that was infinitely malleable by surgical technology. What was important, rather than the nature of the change from body to soul, was the fact that the dualistic principle that held up civilization was maintained: one body should correspond to one sex and only one sex out of the possible two.

The second point of interest raised by these studies is that they show the importance granted to genetics in the determination of biological sex. The enigma that explains the unusual configuration of intersexuals is no longer seen to reside in the gonads or the hormones but in the basic chromosomal traits of every individual and the genes that make him or her up.

Throughout the 1950s, in particular in its applications in experimental agricultural engineering, Spanish genetics began to recover from its dismantling during the post-war period after such a promising republican period.[165] It began to take off once more and recover its position internationally with the work of scientists such as Sánchez-Monge in Madrid and Antoni Prevosti in Catalonia. During the 1960s the first university chairs in genetics were established at the same time as research institutes in agro-engineering and the Laboratory for Cytogenetics, linked in this case to the CSIC, were established.[166]

With respect to sex determination, techniques designed to diagnose chromosomal sex by means of testing cells from inside the mouth as developed by K. L. Moore and his team became known around 1955, as some of the cases above have shown.[167] At the same time, another technique of detecting chromosomal sex by means of the analysis of leucocytes in the blood came on stream. From the start of the 1960s all clinical studies on intersexuality incorporate these techniques as routine.

In addition, from approximately 1959, Spanish researchers became aware of research into those syndromes of intersexuality whose chromosomal aetiology had been established. In particular, research on Klinefelter's syndrome, which was supposed to affect 3 per cent of the population,[168] and Turner's syndrome were received with interest. The first of these syndromes displayed many characteristics similar to those of intersexuality, from genital hypoplasia through to eunuchism and gynaecomastia. Turner's syndrome produced a form of pronounced 'sexual infantilism', especially with undeveloped female gonads, lack of menstruation and secondary sexual characters. In both cases, the therapeutic route embraced hormonal treatment.[169]

This area of research, which really took off from 1959, explained how the hormonal production of the foetal gonads was governed by chromosomal mechanisms. Examples of intersexuality were understood to depend primarily on genetic factors made up by the individual's chromosome map. Errors in the genesis of the gametes or disturbances in embryogenesis could have been provoked by a gene or alterations in genetic processes such as meiosis and could 'conducir a la destrucción de la armonía' (lead to the destruction of harmony),[170] giving rise to multiple forms of intersexuality.

In the 1970s this concentration on the genetic causes of intersexuality would only increase. New resources such as neuroendocrinology that explored the process of feedback between the internal secretions of the gonads, the hypophysis and the hypothalamus were increasingly engaged.[171] Paradoxically, as these new genetic and molecular explanations grew, it appeared that the sexual identity of individuals depended principally on psycho-social factors, or so many argued.[172] Research into the true biological sex, into the dark interstices of the original basis of the organism, became a different project from that of exploring sexual identity or 'gender', a socially forged mechanism projected onto the exterior of bodies. At the same time, the arsenal of surgical techniques expanded and they were employed to realign intersexual identities and, later, were channelled into satisfying the demand of new individuals: those who were deemed transsexuals. In the Spanish case after the democratic Transition, the major thrust of transsexual activism was directed towards making sure that sex-change surgeries were covered by Spain's health service.[173] Sex, understood as a biological truth located in the depths of our being had been subordinated to 'gender', a simulacrum that was infinitely malleable in the operating theatre.

Notes

1 Dreger (1998: 139–66).
2 Oudshoorn (2004: 15–41); Fausto-Sterling (2000: 181–3).
3 Dreger (1998: 181–2).
4 Fausto-Sterling (2000: 63–73).
5 Ibid., p. 44 contrasts the 'Age of Gonads' to an even less flexible 'Age of Conversion' in which doctors found it necessary to identify mixed-sex people at birth and 'convert' them into one sex or the other.
6 Criado y Aguilar (1932a, 1932b).
7 Brachfeld (1932). See the reply by Marañón (1932b).
8 'Con las hormonas y sus interrelaciones se ha explicado todo, hasta lo inexplicable ... Mi actual concepción, experimental, de la doctrina endocrinológica es un tanto escueta, hipercrítica y, si se quiere, ruda ... Pero conforme pasan los años me siento más apegado al caminar tranquilo, y paso a paso, calzados los pies con las botas cautelosas de la responsabilidad y el criticismo, que no al aleteo atrevido de las hipótesis' (With hormones and their interrelation, everything, including the inexplicable, has been explained ... My current experimental conception of the endocrinological doctrine is a little narrow, hypercritical and, if one prefers, a little rough ... But as the years pass I feel happier to tread my path slowly. Step by step, shod with my boots cautious of responsibility and criticism, [I prefer this] to the daring quick step of the hypothesis) (Marañón, 1966c: 569–70).
9 Nash (2000: 689–92); Glick (2005).
10 Castillo Martín (2006: 169–90).
11 Nash (1995: 41–2).
12 Mira (2004: 177–286); Cleminson and Vázquez García (2007: 217–64).
13 Brachfeld reproached Marañón for having expanded overly the remit of the term 'intersexual' from the original use as elaborated by Goldschmidt to include any coexistence between morphological and functional elements of both sexes. Marañón responded that his use of the term was in accordance with existing medical conventions and cited Lipschütz and Bauer to support his case. See Marañón (1932b). The head of the child hygiene service, Muñoyerro Pretel, also justifies Marañón's usage on the basis of Lipschütz and Bauer and insists that

so-called 'pseudo-hermaphrodites' (according to the classification of Klebs) should be considered 'true hermaphrodites' in whom the opposite sex was latent and always alert. See Muñoyerro Pretel (1936a: 95).

14 The acceptance of Marañón's theory of intersexuality as a modern, liberal and progressive notion is noted in many reviews of his main texts. See, for example, Haro García (1929) where it is noted that the 'arcaico criterio de perversidad y monstruosa anormalidad hundida en el pecado, pasa a quedar incluida dentro del campo de la biología' (archaic criterion of perversity and monstrous abnormality sunk in sin, can be included in the field of biology).

15 Marañón (1932b); according to Muñoyerro, from the fourth week the predominant gonad starts to undergo differentiation, a process that continues throughout life. The question is whether the gonad over the life span 'acaba por diferenciarse del todo, o como se sospecha, no termina nunca, quedando en la gónada preponderante siempre vestigios de la gónada del sexo contrario, es decir, en el testículo tejido ovárico, o en el ovario tejido testicular' (becomes completely differentiated or, as we suspect, never finishes and in the predominant gonad traces of the other sex remain, that is, in the testicle there is ovarian tissue and in the ovary testicular material) (Muñoyerro Pretel, 1936a: 95).

16 Marañón (1966b: 135).

17 Marañón and Planelles (1932); Resa (1932); Pardo et al. (1932); Marañón et al. (1932); Marañón (1932a); López Escoriaza and García Orcoyen (1932); Planelles (1932a); Vital Aza (1932); Planelles (1932b); Codina Castellví (1932). Nearly all these articles or notes appeared without the author's initials.

18 Two works in Spanish of the same year that broach these controversies in European endocrinology with respect to hermaphroditism are Bauer (1930: 166–9) and Falta (1930: 420–3). Marañón wrote the prologue to the book by Falta.

19 Marañón (1931a) commented on the means available to determine the presence of male and female hormones in the blood or urine.

20 On the epistemological and institutional conditions that permitted this change in endocrinological thought, see Oudshoorn (1994: 20–2).

21 Ibid., p. 21.

22 Marañón (1966c: 570).

23 Ibid., p. 571.

24 On the theory of original bisexuality as an alternative to the thesis of original asexuality, defended by Steinach, see Verdeguer (1928).

25 Marañón (1931a: 588).

26 On the action of the hormones emanating from the hypophysis see Marañón (1931a: 588, 1966b: 134–5); Planelles (1932: 380); Engle (1938).

27 On the ovarian stimulation of hormonal extracts from the thyroid, see Vital Aza (1932); Grote (1938).

28 Marañón et al. (1932: 898); Marañón (1931a: 588).

29 See Vital Aza (1932).
30 Marañón and Planelles (1932: 129); Kaufman (1938).
31 See n. 139, below on 'progesterone'.
32 On the role of the placenta's hormones in gynaecomastia of lactating men, see Muñoyerro Pretel (1936a: 95).
33 Marañón (1966b: 134–5).
34 'Esta presencia del hormón femenino, positiva en la mayoría de los casos investigados, confirmaría desde el punto de vista químico la teoría actual de la intersexualidad' (This presence of the female hormone, which resulted positive in most cases investigated, would confirm from a chemical point of view the present theory of intersexuality) (Marañón, 1931a: 588).
35 For example, the eunuch of 12 years whose urine 'es positiva en contenido de hormona folicular' (tested positive for follicular hormone) (Marañón, 1931b).
36 For example, the eunuchoid gynaecomast of 17 years who tested positive for 355mm of follicular hormone per litre of urine with no presence of the male hormone (Resa, 1932: 129).
37 Marañón wrote that Frank had found the female hormone in homosexuals, a phenomenon he was researching with Dr Planelles (Marañón, 1931b: 1013). This research was discussed in Marañón (1931a: 588).
38 See, for example, the case of a woman of 39 years who was married to a 'tipo ligeramente afeminado' (slightly effeminate individual) and whose ovary produced virilizing hormones in Marañón (1931c).
39 'Un trabajo de Franck [*sic*] señala, recientemente, el hecho de que hay ciertos pescados que proporcionan cantidades enormes de foliculina, y esto podría explicar, en parte, la frecuencia de la intersexualidad entre los esquimales y otros países ictiófagos' (Some work by Franck [*sic*] has recently shown that there are certain types of fish that contain huge amounts of follicular hormone. This could explain, in part, the frequency of intersexuality amongst Eskimos and other fish-eating countries) (Marañón and Planelles, 1932: 129). Towards the end of the 1930s Frank's thesis was questioned and invalidated by the Amsterdam school, among others (Oudshoorn, 1994: 27).
40 Marañón was aware of this objection. Referring to the finding of follicular hormone in the urine of homosexuals, he noted that 'no categorical significance' could be read into the findings (Marañón, 1932a: 280).
41 Oudshoorn (1994: 24–7).
42 This antagonism found explicit formulation in Marañón's work and that of his co-workers: 'lo cierto para el desarrollo evolutivo de la sexualidad, es el claro antagonismo entre el ovario y el testículo' (what is certain in the evolutionary development of sexuality is a clear antagonism between ovary and testicle) (Verdeguer, 1928: 210). Some authors, even though they accepted Zondek's findings, persisted in interpreting them in a way that defended the antagonism

thesis. For example, the increase in follicular hormone in male castrates, even though large amounts of the hormone were found in non-castrates, was seen to prove the 'antagonismo normal entre las dos secreciones, macho y hembra' (normal antagonism between the two secretions, male and female) (Quental, 1938).

43 The four cases included photographs that showed full frontal images of the genitalia of the patients. We do not consider here the cases of intersexuality discussed in Marañón's circle, or that of the recruit in 1881 who tried to pass himself off as a woman to avoid military service (discussed in Gimeno Cabañas, 1931).

44 Muñoyerro Pretel (1931).

45 'No tiene, pues, nada de extraño que se consideren al hermafroditismo y pseudohermafroditismo en un mismo grupo, en el que sólo existen variaciones de grado, es decir, que todos son verdaderos hermafroditas' (It is not in the least strange to consider hermaphroditism and pseudo-hermaphroditism in the same group, in which there exist only differences of extent, that is to say, all are true hermaphrodites) (ibid., p. 506).

46 Mas Collellmir (1932: 14). In order to buttress his claims for true hermaphroditism, Mas Collellmir declared that there were three types of true hermaphroditism: bilateral, unilateral and lateral. These classifications coincided with those of Klebs (Dreger, 1998: 145).

47 Pons Tortella and Gállego Berenguer (1934).

48 The shorter oral version, presented to the Pediatric Society of Madrid, was published twice: Muñoyerro Pretel (1936b, 1936c).

49 Muñoyerro Pretel (1936a: 95).

50 See, for example, Thoinot (1928: 115).

51 Exploratory surgery or laparotomy was invented in France in the 1890s and only began to be used widely in France and Britain from the 1920s onwards (Dreger, 1998: 93). We have not found evidence of the use of this surgical technique in the medical reports of the 1930s in Spain.

52 See the case of a girl of 15, a hermaphrodite as a result of the functional insufficiency of the female gonad whose 'estudio no fue lo suficientemente completo, por las condiciones de la enferma, al pertenecer a clientela particular' (study had not been sufficiently complete because of her particular condition as she was a private client) (López Escoriaza and García Orcoyen, 1932: 520).

53 See, for example, Richards (1998); Polo Blanco (2005), part of which appears as Polo Blanco (2006) and the various chapters in Huertas and Ortiz (1997).

54 Juliá (2004: 287–97).

55 Roca i Girona (1996); Richards (1998: 24–66); Juliá (2004: 295–6). Polo (2006: 27) uses the term 'palingenesia' to refer to this type of new birth.

56 'Para que España vuelva a ser es necesario que la Institución Libre de Enseñanza no sea' (In order that Spain comes to be once more, it is vital that the Institución Libre de Enseñanza does not exist)

(F. Martín-Sánchez Juliá, 'Origen, ideas e historia de la Institución Libre de Enseñanza' (1940), cited in Juliá (2004: 293).

57 This expression, coined by Laín Entralgo, names one of the main tasks of this undertaking of 'depuración'. See Claret Miranda (2006).

58 The 'purification' of the Faculty of Medicine at the Central University of Madrid is analysed by Otero Carvajal (2006). In Catalonia, these dismissals affected about 10% of doctors in 1936. See Reventós et al. (1991: 28–9).

59 Marañón, professor of endocrinology at the Central University of Madrid, left for Paris shortly after the start of the war from where he actively favoured the military uprising. Despite this, in 1937 all his assets were seized by the regime. In 1939, pressed by Leonardo de la Peña, professor of urology at the same university, the Tribunal of Political Responsibilities opened up another case against him. He was cleared in 1944 after having returned to Spain the previous year (Otero Carvajal, 2006: 103).

60 Di Febo (2006: 217–18); Richards (1998: 55–8); Mira (2004: 287–9).

61 On the re-establishment of these rigid gender divisions, see Folguera Crespo (1997: 528–30); García-Nieto París (2000: 724–6).

62 'Racial' here is used not in the sense of 'ethnic racism' but with the meaning of 'biological racism'. On the distinction, see Foucault (1997: 70–2).

63 Polo Blanco (2006: 503). On the objective of 40 million Spaniards, see Di Febo (2003: 32–3).

64 Polo Blanco (2003: 112–15) and Clavero Núñez (1953: 443).

65 This task was given principally to the women of Falange's Sección Femenina. See Richmond (2003: 67–70).

66 'El matrimonio debemos considerarlo como la meta por naturaleza, que representa biológicamente una equilibración ... Mi hipótesis es que cada sexo por separado es física y espiritualmente un organismo en equilibrio funcional sólo aparente, que no subviene por sí a todas sus necesidades o apetencias' (We should consider matrimony to be the natural aim, which represents biological equilibrium ... My hypothesis is that each separate sex is physically and spiritually in a merely apparent functional equilibrium, which does not provide for all its needs and desires on its own) (González Galván, 1940).

67 'La maternidad es, en una palabra, la expresión funcional más caracterizada de la constitución de la mujer' (Maternity is, in a word, the functional expression that is most characteristic of woman) (Clavero Núñez, 1953: 48). On maternity in the post-war period, see Polo Blanco (2003); Roca i Girona (1996, 2003).

68 One of the most severe critics of the theories of Marañón was without doubt Vallejo Nágera, who considered that endocrinology was a discipline that had practically no use for psychology. He believed that the concept of intersexuality laid waste the differences between the sexes and signified an attack on Catholic morality by liberal medicine. On this idea, see Richards (2004: 841). It should be recalled that in the post-war purge Vallejo Nágera acted as a witness against Marañón.

69 On the promotion of this line of research, see Polo Blanco (2006: 195–7).
70 The two great authorities in the struggle against sterility in the gynaecological field during the Franco period were undoubtedly Antonio Clavero Núñez (1903–62), a maternologist and editor of the *Revista de Obstetricia y Ginecología* and José Botella Llusiá (1912–2002), professor of obstetrics and gynaecology at Zaragoza. Botella published in 1946 a *Tratado de ginecología* which had achieved fourteen edns by 1993. From 1971 he revised the work with the help of the son of Antonio Clavero, José Antonio Clavero Núñez, the current chair of the discipline at the Complutense University, Madrid.
71 Polo Blanco (2006: 237–9).
72 'cada una de ellas [las glándulas] tiene una misión propia, especialísima ... las genitales en los fenómenos del sexo, las suprarrenales en el tipo orgánico ... Las gónadas forman la piedra angular de todo el mecanismo sexual. Son específicos [sic] para el género; ovarios para la mujer, testículos en el hombre' (each one of them [the glands] has a very special mission of its own ... the genital glands in the sex phenomena and the suprarenal in the organic sphere ... The gonads form the corner-stone of the sexual mechanism. They are gender-specific – ovaries for women and testicles in men) (González Galván, 1940: 72). The use of gender here, rather than denoting what we now understand by the term, appears to be merely in order not to repeat 'sex' in the text.
73 Bañuelos (1941: 860–7); Botella Llusiá (1941, 1946); Vital Aza (1941); Peña Regidor (1941); Clavero Núñez (1942); Díaz del Villar (1943); Juaristi (1943); Villarreal Casas (1946); Cónill Montobbio (1946).
74 Cifuentes Delatte (1941) is a completely up-to-date exposition of the question.
75 Bañuelos (1941: 863–4).
76 The conservative, pro-regime gynaecologist Vital Aza had considered that hormone therapy was only justified in order to restore 'con una plena eficacia de la biología de la mujer adulta el feliz cumplimiento de sus aptitudes generatrices' (the happy fulfilment of woman's reproductive aptitudes with the maximum level of efficiency that the biology of an adult woman can provide). This form of prophylaxis of sterility would 'salvar en la niña de hoy, la madre de mañana' (save in today's young girl, the woman of tomorrow) (Vital Aza, 1932: 381).
77 Regarding the eunuch: 'no presenta aquél los rasgos viriles característicos de valor, capacidad de crítica, etcétera' (he does not present the characteristic virile traits of bravery, critical faculty, etc.) (Bañuelos, 1941: 861).
78 On women in the climacteric, Bañuelos indicates that they show 'modificaciones del carácter que tiende a hacerse más o menos viriloide o irritable ... con gran frecuencia todas las molestias de las mujeres en este periodo están impregnadas de un sello francamente neurótico' (character changes tending towards the more virile and

more irritable ... frequently all the sufferings of women during this period are impregnated with a frankly neurotic aspect) (Bañuelos, 1941: 865).
79 Cifuentes Delatte (1941: 15). Similarly, Cónill Montobbio, professor of obstetrics and gynaecology at the University of Madrid, wrote that 'hay ciertas prevenciones acerca del empleo, en la mujer, de la hormona masculina ... por los indicios viriloides (hirsutismo, involución de mamas, hipertrofia de clítoris, etc.) que produce y que hacen temer la atrofia ovárica tal vez irreparable' (there are certain precautions to be taken in the administering of the male hormone in women ... because of the viriloid tendencies it produces (hirsutism, reduction in breast size, hypertrophy of the clitoris, etc.) and which may entail perhaps irreparable ovarian atrophy) (Cónill Montobbio, 1946: 95). The collective noun 'androgens' incorporated androsterone and testosterone. See Oudshoorn (2004: 36).
80 Cifuentes Delatte (1941: 15).
81 Some doctors cannot deny the lack of sex-specificity of the hormones but try to relativize it by saying that in any individual some hormones dominate others and this influences decisively sex determination. See Simarro Puig et al. (1942). The impossibility of avoiding the lack of sexual specificity of the hormones proves that certain 'facts' are not infinitely malleable and cannot always be reinterpreted by language. They offer a form of resistance that confirms their epistemological objectivity, as Searle might say, as well as being 'ontologically objective'. They are not 'mere' social constructs. See Searle (1996); Hacking (1999).
82 Simarro Puig et al. (1942: 39) note that 'las delicadas situaciones de índole moral y social a que tales alteraciones dan lugar explican la obligada discreción y silenciamiento de tales hechos' (the delicate situations of a moral nature that such alterations give rise to explain the obligatory discretion and silence that covers such facts).
83 Simarro Puig et al. (ibid.) write 'de acuerdo con los conocimientos actuales sobre endocrinología, permiten invalidar el hermafroditismo verdadero, por lo menos para un número no despreciable de tales observaciones' (following current knowledge in endocrinology, it is possible to invalidate true hermaphroditism, at least for a not inconsiderable number of observations). Roda et al. (1944: 170) record that for 'true hermaphroditism' to be the case 'requeriría que el sujeto hermafrodita fuera capaz de producir al mismo tiempo óvulos y espermatozoides, lo que hasta la fecha no se ha comprobado ni en un solo caso' (the hermaphrodite subject would have to be capable of producing sperm and eggs at the same time, something which, to date, has not been recorded in a single case).
84 Simarro Puig et al. (1942: 39).
85 On Louis Ombrédanne, see Androutsos (2003). Ombrédanne believed that sex determination in doubtful cases depended on not one single decisive factor but on what he termed the 'sexual balance'

of the subject, which would include the relative preponderance of primary, secondary and tertiary sexual characters. On this concept, see Cónill Montobbio (1946: 317).
86 Simarro Puig et al. (1942: 39) believed that sex determination and the real sex of the subject corresponded to the 'active' gonad, capable of procreation.
87 Ibid., p. 40.
88 Ibid.
89 Ibid., p. 41. If this operation had gone ahead, it would have unintentionally been the first sex-change operation to take place in Europe that we know of.
90 Ibid., p. 42.
91 Ibid.
92 Ibid., p. 41.
93 In 1929 the Holy Office had responded in the negative to the question as to whether it was legitimate to obtain semen by means of masturbation in order to analyse it clinically. The sample could only be obtained by massaging the seminal ducts, spontaneous ejaculation (nocturnal pollution) or by collection from the vagina after coitus with the husband. It was this method that appears to have been recommended by doctors who worked on fertility in the maternity houses, according to Polo Blanco (2006: 194–5). Obviously, in this case, this method was not possible since the subject was single.
94 Simarro Puig et al., p. 39.
95 Ibid., p. 42.
96 Piulachs (1943: 212).
97 'Pozzi ha puesto en duda la existencia del hermafroditismo verdadero, que, sin embargo, es confirmado por muchos autores. En las obras de Lagos García [*Las Deformidades de la sexualidad humana*, Buenos Aires, 1925] y Ombrédanne se recopilan numerosos casos' (Pozzi has cast doubt on the existence of true hermaphroditism, which, however, is confirmed by many authors. In the work of Lagos García [*Las Deformidades de la sexualidad humana*, Buenos Aires, 1925] and Ombrédanne numerous cases are presented) (Piulachs, 1943: 211).
98 Ibid., p. 212.
99 On the relationship in the male between virility and number of children born, see Polo Blanco (2003: 95).
100 We have not consulted the case of 'true hermaphroditism' in a boy of 11, published in 1945 by Drs García Portela and Fernández Ruiz in the *Revista de Obstetricia y Ginecología* (see Cónill Montobbio, 1946: 316).
101 Roda et al. (1944: 165).
102 Ibid., p. 166.
103 Ibid., p. 168.
104 The authors also advocated the reading of the work by Allen on sex and the internal secretions and Young on genital abnormalities (Roda et al., 1944: 171). By the time this article appeared (May 1944)

Marañón had already returned to Spain and was awaiting his clearing by the Tribunal for Political Responsibilities. As noted, this took place Nov. 1944 (Otero Carvajal, 2006: 103). The praising of Marañón's work, therefore, was part of his rehabilitation.
105 This possibility had been remarked upon by Marañón. Abnormal intersexualities could appear when the chromosomal mix results in an excess in virile sexuality in the female or an excess in female sexuality in the male. This excess could be corrected by the action of the internal secretions. Likewise, a normal individual could lose their normality through the defective action of the internal secretions. See Marañón (1950: 538).
106 Roda et al. (1944: 170). A manual from the period, after discussing the masculinizing effects of testosterone and the feminizing effects of oestrogen and before showing the reverse dynamic and the lack of sexual specificity of the hormones, states: 'la contraprueba sugiere importantes meditaciones' (proof of the reverse [phenomenon] suggests some important issues) (Cónill Montobbio, 1946: 95).
107 See Roda et al. (1944: 170). This more restrictive criterion can be compared with a more comprehensive outlook from Marañón during the same period: 'el examen histológico de las gónadas permite distinguir hermafroditismo de seudohermafroditismo, aunque la distinción es de matiz, pues en el tejido gonadal del seudohermafrodita hay siempre elementos del otro sexo ... En resumen, entre el hermafroditismo y el seudohermafroditismo no existen más que diferencias de grado a veces difíciles de establecer' (the histological examination of the gonads allows us to distinguish hermaphroditism from pseudo-hermaphroditism, although the distinction is a fine one, because in the gonadal tissue of the pseudohermaphrodite there are always elements of the other sex ... To sum up, between hermaphroditism and pseudo-hermaphroditism there only exist differences in extent, which are difficult to determine) (Marañón, 1950: 539). An intermediate position is represented by Cónill Montobbio, who maintains that 'true hermaphroditism' does occur in the human species, even functionally, and records cases of production of both eggs and sperm and of ovo-testis. What is impossible is 'total or complete hermaphroditism' (Cónill Montobbio, 1946: 315).
108 Roda et al. (1944: 171).
109 Ibid.
110 Ibid.
111 Ibid., p. 170.
112 In some texts from the period, however, the impression is given that if the 'true sex' is revealed to the male pseudo-hermaphrodite, he 'repudia su indumentaria, su nombre y todo su plan femeninos' (refuses his female dress, name and whole female trajectory) (Cónill Montobbio, 1946: 316).
113 Roda et al. (1944: 170).

114 'Desde el punto de vista terapéutico se ha modificado modernamente el precepto clásico de defender, o dicho de otro modo, pretender restaurar el sexo vergonzante del seudohermafrodita. El plan que con discreción y secreto profesional realizó Crossen (1939) en un andrógino nos parece más atento a la dignidad personal que el notición de un cambio de sexo' (From the therapeutic point of view, recently the classic position of defending the restoration of the hidden [literally, shameful] sex of the pseudo-hermaphrodite has been revised. The strategy that Crossen (1939) developed with professional discretion and confidentiality with an androgyne seems to us to attend more to personal dignity than does the big news item of a change of sex) (Cónill Montobbio, 1946: 316–17). The same author cites an example of this method and indicates that 'se sofocó el verdadero sexo en beneficio del que el enfermo ejercía ya' (the true sex was suffocated in favour of the sex that the patient exercised then) (317). As can be seen, the 'personal dignity' of the person included avoiding social scandal, which the change of sex in a rigidly gendered society would have caused.

115 Balthazard (1947: 462–3) wrote that 'neutrals' or those of 'incomplete sex' would attract the interest of 'pederasts'.

116 Botella Llusiá studied medicine at the Central University, Madrid, and graduated in 1934. He later studied in Germany and Austria, obtaining his first chair in obstetrics and gynaecology in Zaragoza in 1944 and in Madrid in 1947. In 1950 he was accepted as a member of the Royal Academy of Medicine. He was made President of the Academy in 1986 and in 1956 a member of the Higher Council for Scientific Research (CSIC). He joined Opus Dei and was a member of the Council of State under Franco. Between 1968 and 1972 he was chancellor of the Complutense University of Madrid, a post which he resigned.

117 Botella Llusiá (1943, 1944). At the start of the 1970s Botella Llusiá began to favour methods of birth control but only if they were compatible with the Catholic faith (Botella Llusiá, 1971: 80) After the transition to democracy, he would be a major opponent not only to any law on abortion but also to pharmaceutical contraceptives.

118 Botella Llusiá (1941, 1946).

119 On this shift, which resulted in a struggle between more 'liberal' and integrist Falangists within National Catholic discourse, see Juliá (2004: 355–408).

120 In 1945 the Spaniards' Charter and the Law on National Referendum were passed. Shortly afterwards, the Law of Succession was passed in 1947. This made the former two laws part of the 'Fundamental Laws' of the regime together with the Labour Charter. The Law of the Principles of the Movement, presented by the regime's acolytes as a Magna Carta of 'organic democracy' was approved in 1958. See Díaz (1974: 53–5).

121 This is the socio-economic context of the law of 22 July 1961, which gave women the same rights as men to exercise any kind of political

and professional activity, save the military, navy and certain classes of 'dangerous' work. See Folguera Crespo (1997: 543); García-Nieto París (2000: 730–2). On Mercedes Formica, who reviewed de Beauvoir's *The Second Sex* in 1950, see Nielfa Cristóbal (2003: 275–8).
122 Botella Llusiá (1953).
123 Ibid., p. 367.
124 Ibid., p. 369.
125 See the essay written by the ex-policeman Carlavilla del Barrio (1956: 9). The first part of the book is indeed called 'Sodomy and Communism' (pp. 11–67). On the homophobia of this period of Francoism see Mira (2004: 316–20).
126 Carlavilla devotes an entire chapter to Marañón and considers him to be the head of the 'escuela científica sodomizante' (sodomitic scientific school) in Spain (Carlavilla del Barrio, 1956: 77–124).
127 In 1954 the regime approved an amendment to the Ley de Vagos y Maleantes drawn up during the republican period to repress homosexuality and in 1970 the Law on Social Dangerousness and Social Rehabilitation was adopted (Mira, 2004: 320–8).
128 Botella Llusiá (1953: 371).
129 Ibid.
130 The importance of the 'dignity' and the welfare of the patient are constantly reiterated. The author writes that it is necessary to achieve 'la satisfacción de su clasificación en un sexo conforme con sus instintos' (the satisfaction of his classification in a sex that conforms with his instincts) (ibid., p. 372); 'optar por aquél que pueda reportar mayor bien al paciente' (to opt for that [sex] that will entail the greatest well-being for the patient) (ibid.); and to elect the sex 'que ellos creen que es el suyo y bajo el que pueden conseguir la dignidad y la felicidad' (that they believe is their own and with which they can achieve dignity and happiness) (p. 373).
131 Ibid., p. 372. The discussion on pseudo-hermaphroditism in the chapter on the 'anomalies and sexual perversions' in medico-legal accounts of the period, as well as the concern felt in the case of scandals, can be seen in, for example, Simonin (1962: 414). Here, it was noted that androgynous pseudo-hermaphrodites 'pueden tratar de actuar como varón y provocar escándalos, pero en la mayoría de los casos la erección no es suficientemente libre para penetrar una vagina' (could try to act as males and provoke scandals, but in the majority of cases their erection is not sufficiently developed to be able to penetrate the vagina).
132 Previous chapters have analysed this conundrum. See Dreger (1998: 119–38).
133 Botella Llusiá (1953: 371) wrote 'ante un manifiesto error de sexo, nada mejor que rectificarlo' (when faced with an evident error of sex, it is best to rectify it). In practice, however, Llusiá would be guided by the social sex. Here we see an example of old paradigms lingering on.
134 Many years later, Botella Llusiá believed he had found this basis in sociobiology. While the passive and active character of the sexes

resulted from a biological basis, cultural influences could alter them 'artificially'. See Botella Llusiá (1983: 7)

135 'La Biología tiene obligación de investigar hasta lo más hondo los problemas del sexo, pero yo creo que la Moral y el Derecho deberían detenerse ante las puertas de este misterio. Mientras no sepamos por qué se determina el sexo, mientras que no nos sea dado definir hasta sus últimas consecuencias la esencia de la sexualidad, es preferible aceptar que las cosas son, en último término, lo que parece que son' (Biology has the duty to investigate as profoundly as possible sexual problems but I believe that morality and law should refrain from opening the doors of this mystery. Until we know what determines sex, until we are to define the essence of sexuality with all its consequences, it is preferable to accept that things are, in the last analysis, how they appear to be) (Botella Llusiá, 1953: 373).

136 On the rise of the psychological challenge to biological models see Rose (1999: 157–81). On the attempt in Spain to build a behaviourist form of psychotherapy which would fuse Christian and Freudian perspectivas, see González Duro (1996: vol. 3, 318–23).

137 Carlavilla places political figures such as Manuel Azaña and Diego Martínez Barrio amongst biological 'eunuchoids' and relates this quality to supposed homosexual and conspiratorial tendencies. See, respectively for Azaña and Martínez Barrio, Carlavilla del Barrio (1956: 137–61, 163–204).

138 The legend of 'La Pastora' was created by the journalist Enrique Rubio, who published in the sensationalist *El Caso* at the end of the 1950s a series of articles on the activities of Teresa, the anarchist 'hermaphrodite maquis'. The legend became well known in 1978 when the writer Manuel Villar Raso, basing his account apparently on an autobiographical sketch by Teresa herself, published her biography. Teresa Pla and the former prison guard Marino Vinuesa, the person who helped La Pastora after she left prison and who worked on her autobiography, accused Villar Raso of plagiarism and falsification of true events. See the novel by Villar Raso (1978) and the interview with Teresa Pla by M. Alberola (1988).

139 Botella Llusiá and Nogales (1952). We have consulted the review of this case written by Señor (1953). In addition, Botella Llusiá presented his ideas on the internal secretions of women in Botella Llusiá (1955). Here, he maintained that there were two types of specifically female hormones, oestrogen or the hormone 'of femininity', linked to the 'attraction of the male' and progesterone, the hormone of maternity, which protected the 'mechanism of fertilization'. In addition, he recognized that the presence of androgens in the urine of normal women was not an index of intersexuality or any remains of original hermaphroditism. It was merely a 'physiological fact' (pp. 106–7).

140 Botella Llusiá, *Dos casos de hermafroditismo verdadero (Conferencia en la Real Academia Nacional de Medicina, día 24 de enero de 1956)* (Madrid: Imprenta de José Luis Cosano, 1956). He presented a new case of true hermaphroditism in Botella Llusiá (1966).

141 Botella Llusiá (1956: 4).
142 Ibid., p. 13.
143 But in fact Botella Llusiá does not practise this kind of analysis in the two cases mentioned. He merely enters into conjecture. This technique is perfected at the start of the 1950s. See Editorial (1955).
144 Botella Llusiá (1956: 17–19).
145 Botella writes 'A nuestro modo de ver, la líbido de estos sujetos está determinada por el concepto que ellos tienen de su propia sexualidad' (In our view, the libido of these subjects is determined by the concept that they have of their own sexuality) (ibid., p. 21).
146 Ibid., p. 22.
147 Ibid.
148 Ibid.
149 'Mientras estos aspectos no estén totalmente esclarecidos, debemos admitir provisionalmente que todo individuo es hombre o mujer, más que según la naturaleza de su propia gónada, según su apariencia externa y según sus impulsos, y debemos elegir para estos sujetos aquel sexo que ellos creen que es el suyo y dentro del cual pueden conseguir la felicidad y la dignidad dentro de la sociedad en la que viven' (Until these issues are clarified completely, we should accept provisionally that all individuals are either men or women, rather than in accordance with the nature of their own gonads, in accordance with the external appearance and their impulses. We should elect for these subjects the sex that they believe is their own, within which they can achieve happiness and dignity in the society in which they live) (ibid., p. 23).
150 A prestigious manual on hormonal therapy coincided with Botella's proposal in this sense: 'la pregunta si en general está indicado tratamiento, depende de la elección de sexo que se haya hecho tras el nacimiento y del desarrollo individual. El hallazgo gonadal y cromosomático, que muchas veces no coincide con el hábito externo y el desarrollo psíquico, no justifica el intento de hacer un "cambio de sexo" terapéutico' (the question as to whether in general treatment is to be followed depends on the election of sex that has been made after birth and individual development. The gonadal and chromosomal evidence, which often does not coincide with the external appearance and psychic development, does not justify the attempt to implement a therapeutical 'change of sex') (Ufer, 1960: 106).
151 Fontes Gil et al. (1960).
152 Ortega Núñez et al. (1960).
153 Cañadell and Planas Guasch (1961). We have not examined the three clincial cases they cited of true hermaphroditism in Spain that were published between 1957 and 1961. These materials are: M. Álvarez Coca, M. Aguirre, G. Gobeo and F. Ferrán, 'Hermafroditismo verdadero alternante', *Revista Ibérica de Endocrinología* (1957); A. Aznar Reig, 'Hermafroditismo verdadero alternante: caso clínico', presentation given to the I European Colloquium on

Endocrinology (1961) and M. Fernández Fernandes, 'Alteraciones de la sexualidad o disgenopatías gonadales', *Revista Ibérica de Endocrinología* (1961).
154 Fontes Gil et al. (1960: 314). On the dangers that this attitude could entail for certain classes of intersexuality, apart from the moral paternalism implied, see Dreger (1998: 188–96).
155 Cañadell and Planas Guasch (1961: 335).
156 Ibid., p. 343.
157 On the work of Money and his team, see Fausto-Sterling (2000: 46, 63–77); Dreger (1998: 181–2). The article referred to by the Spanish authors was Money et al. (1955).
158 Cañadell and Planas Guasch (1961: 339).
159 Ibid., p. 343.
160 Ibid. On the changes of the Francoist state towards a welfare state model with respect to health, see Varela and Álvarez-Uría (1989: 61–72).
161 Cañadell and Planas Guasch (1961: 343).
162 Ibid.
163 The term 'transsexual', or at least 'psychopathia transexualis', was introduced by an article published in 1949 by D. O. Cauldwell but the concept 'transsexualism' only gained a hold in the discipline in the early 1950s when it was differentiated from transvestism. In 1974 the term 'gender dysphoria' was coined as a broader phenomenon than transsexualism, referring to all those individuals who have experienced discomfort with their biological sex and who seek a new sex (Hausman, 1992).
164 The first to undertake a large number of sex-change operations were the medical teams at the Johns Hopkins University and they were the same teams that had specialized in intersexual surgery years before, according to Mejía (2006: 110–11). In Spain, Botella Llusiá classified transsexualism as 'voluntary intersexualism'. The former 'añade un carácter iatrogénico a lo que antes era primitivo' (adds an iatrogenic aspect to what was foundational before). See Botella Llusiá (1987: 5).
165 Pinar (2002).
166 Ibid., pp. 139–40.
167 Editorial (1955).
168 Morer Fargas (1962: 168). This paper was originally given at the V Seminar on Endocrinology on 27 Dec. 1961. The statistics discussed by Fausto-Sterling (2000: 53), however, show a frequency of 0.0922 for every 100 live births.
169 Morer Fargas (1962: 167–9).
170 Ibid., p. 166.
171 Pallardo Sánchez et al. (1972).
172 This disassociation is remarked upon in the translation of the 1975 monograph on hermaphroditism written by the Johns Hopkins medics Howard Jones and William Scott. Five biological criteria are proposed for the determination of sex (chromosomal type, gonadal

structure, external genitalia, internal genitalia and hormonal state) as well as two psychological ones (sex of rearing and gender role of the individual). The notion of transsexualism as distinct from hermaphroditism is introduced and the former is understood primarily as a 'psychiatric problem' distinct from hermaphroditism per se. See Jones and Scott (1958: 46).

173 Mejía (2006: 31–8).

Chapter 6
Conclusions

The preceding chapters, admittedly with significant chronological gaps, have traced the evolution of scientific discourse on the hermaphrodite in Spain from 1560 to 1960. In order to do so, we have applied the insights of authors such as Lynda Birke, Alice D. Dreger, Nelly Oudshoorn and Anne Fausto-Sterling in an attempt to transcend any conceptual dichotomies between sex and gender, an issue recently addressed by feminist accounts.

Scientific discourse on biological sex over the time period analysed here has been traced with reference to the changing history of medical technologies – a pragmatist perspective has been adopted here[1] – and in the light of changing political, social and gender-related backdrops. In this way, the body, sex, sexual orientation and gender itself are understood not as independent variables but as social institutions, which are made up in accordance with their own different historical conditions. These varied historical configurations (heterogeneous groups of discourses, artefacts, technologies, forms of political and social organization) do not constitute different versions in themselves of the 'hermaphrodite' as an entity that was unchanged and unaffected by historical discourse, as if it were a natural object.

Instead, in each of these historical 'assemblages' the shape of the object in question takes on a different form. The 'hidden hermaphrodite' present in the literature of marvels and wonders of nature or in the anatomical treatises of the sixteenth century is as different a personage from the 'apparent hermaphrodite' profiled by medico-legal texts in 1850 as it is from the 'intersexual' analysed by endocrinology in 1930. These differences do not take on a kaleidoscopic form or assume the form of icebergs isolated from one another.[2] One of the main tasks of this volume has been to trace the similarities, the lineage, between different expressions of hermaphroditism while accounting for their differences. The overall picture, therefore, is one of interconnected shifts rather than a succession of discontinuous blocs. Alongside some elements that are present throughout the whole historical period discussed (for example, the association between hermaphroditism and sexual deviance) there are other characteristics that fade away rapidly, such as the medical interest in hypospadias at the beginning of the twentieth century. Our understanding of hermaphroditism is distanced also from those visions that attempt to capture in any given period the dominant theoretical framework in which hermaphroditism was understood. We have not tried to see definitively whether the 'one-sex' or the 'two-sex' model held sway at any particular moment, as some historians have, following the work of Thomas Laqueur. In the Spanish case, each and every historical configuration that we have identified displays an unstable heterogeneous variety of ways of seeing and speaking about the subject, which are irreducible to a single 'ideal type'. The terms we have employed, such as 'sex as status' or the period of 'true sex' do not represent unitary cosmovisions but dispersed understandings that are constantly 'resignified' as part of new contexts.[3]

Examples of this kind of pluralism can be located in each period examined here. During the period of 'sex as status', there was no homogeneous *Weltanschauung* but a diversity of discursive and non-discursive regimes configured by the triple distinction between the experience of *mirabilis, magicus* and *miraculosus*. This heterogeneity was amplified in the transition period of the eighteenth century during which the attempts to demystify (for example, the enlightened critique of sex changes) coexisted with more traditional thought. During the period characterized most evidently by the notion of the 'true sex' (1820–80) also apparent was an irreducible discursive diversity in which concepts from legal medicine, general pathology and the first texts from sexual hygiene overlapped. Around 1910–30 the picture is extraordinarily diverse and four main ideas are in competition: the tendency to emphasize the problem of hypospadias to the detriment of full hermaphroditism by relying on degenerationist thought; the reference to 'pseudo-hermaphroditism' and to the gonadal criteria as a key to determining true sex; the first endocrinological accounts, which foregrounded the importance of the hormones, with Marañón's theory of 'intersexuality' gaining ground. From the 1930s up to the 1950s there was both critique and acceptance of Marañón's theory, which in turn brought the gonadal theories of Klebs and Pozzi back into the frame with the assistance of the ideas of Ombrédanne. Lastly, from the mid-1950s onwards we note a process of polarization whereby research on the biological determination of sex shifts to the chromosomal plane and is disassociated from identifying sexual identity, which in turn becomes more firmly lodged in the realm of psychology and psychiatry.

In this changing panorama it is true to say that those voices that are most clearly vocalized are those of the specialists in charge of objectivizing hermaphroditism – doctors, legal experts, theologians, journalists, poets and painters – with those designated hermaphrodites often reduced to silence. However, within the many accounts of the 'abject lives'[4] related in the descriptions of *mirabilia*, 'relaciones de sucesos', *exempla anathomica*, legal or clinical cases, the resistance, the power of self-affirmation and the creativity of the subject is often heard.

It is true that, in the Spanish case, we have not been fortunate enough to find anything like the memoirs of a figure like Herculine Barbin, although the first-person accounts of individuals such as Elena de Céspedes or Catalina de Erauso are equally revealing and important in their own way, as are the intrepid and historically important accounts of legendary figures such as Brígida del Río at the time of the Spanish Armada, Reyes Carrasco and the anarchist guerrilla Teresa Pla. Other figures such as the Madrid servant girl who initially resisted the examinations of Dr Ulibarri in 1860, the bemused husband and wife who related their sexual practices to Dr Pascual in 1920, the multitude of humble folk, mainly farm workers, whose equivocal bodies were used in order to try out the nosological frameworks of Marc, Klebs, Pozzi or Marañón – they were all studied with the same interest and enthusiasm in the medical academies and surgeries or contemplated with fascination or ridicule in the fairgrounds and village squares as other more well-known cases in other countries.

It cannot be denied that the ways in which each period codified the differences between the sexes conditioned the relationship that those affected had with their own bodies. Consider, for example, the reaction of the father of María Muñoz, the nun from the convent in Úbeda who changed into a man in 1617. The father, on receiving the news, swiftly exchanged his stupor for unbridled glee. The sex change allowed him to transmit the whole of his fortune to his new-found son. The understanding of sex as a sign of status, that is, as bestowing rights and prerogatives rather than a constituting a biological 'fact', and of the body as a malleable structure, allows us to comprehend a reaction that seems bizarre to us today. An even more eloquent case is that of Elena de Céspedes. By uniting her erudition with surgical dexterity, this Andalusian slave was capable of presenting herself as a 'hermaphrodite' and thus cast doubt on the arguments of her accusers. By relying on the Hippocratic-Galenic one-sex model and through the manipulation of her own body, she managed to confound the doctors that examined her.

However, with the consolidation of legal medicine and its desire to discover the 'true sex' of the individual, it becomes more difficult to determine the point of view of those affected and their conditioning by the knowledge of the period. It will be recalled that the rules drawn up by Marc at the beginning of the nineteenth century urged caution or disbelief when listening to the life story of 'hermaphrodites'. The latter went or were taken to the doctors for a variety of reasons. They may have been children or newly born babies suspected of having been assigned the incorrect sex, a moment when the enlightened doctor would pronounce against the 'ignorant' midwife. Or it may have been a case of illness such as a hernia, which became full of significance once they were examined. Individuals may have sought medical advice, tired of the ridicule to which they were subject by others because of their 'ambiguous' bodies, or may have been concerned about their lack of competence in the marital bed. Others were made known during the medical examination of army recruits. Baptizing a boy as a girl was one way of avoiding military service and examples of this are many, including the case of 1881, or the more collective practice in the village of Los Lagares where, at the beginning of the twentieth century, peasants registered their newly born boys as girls in order to avoid forced military recruitment and the consequent loss of farmhands.

The majority of these cases came from small villages or towns, from the interior rather than from the coastal areas and overwhelmingly from peasant environments, where it was not the custom to listen to one's body, to display it to others or to talk openly about sexual life. This accounts for the resistance towards medical examinations in individuals not used to doctors' surgeries or medical terminology in the late nineteenth and early twentieth centuries.[5] This bodily culture, an index of carelessness and even savagery for the urban elites, multiplied the possible appearance of incorrectly identified children or wives who hid their true identities. Usually, these ambiguous figures – women undertaking hard agricultural labour, unmarried males confined to the parents' home – were well integrated into community life in spite of their sexual peculiarities, to the extent that they could spend their whole lives without being 'discovered'. On the other hand, it is quite possible that there were more cases of interest that never reached the medical press, being assessed in the private clinics attended by Spain's richer classes.

The silence of the patient began to thaw in Spain in the 1920s. From this date onwards, the sexual proclivities of the individual took on an importance that was reflected in the final diagnosis. This interest in sexual preferences and activities mushroomed in the 1950s when the importance of the biological determination of sex (by means of chromosomal analysis) was supplanted by an analysis of the psycho-social process of learning. The reports by Botella Llusiá are indicative of this shift.

In the process of interaction between doctor and patient the knowledge held by the former conditioned the experience of the patient but only to a certain degree. We have already discussed the existence of specific bodily cultures that resisted examination and produced different knowledge from that of the doctors. In addition, as patients' voices were heard more clearly, the categories used by specialists evolved in accordance with the information supplied by their subjects of examination. The tendency towards individualized diagnoses, which was very prominent in the 1870s, when the limitations of mere visual analysis of the external genitalia and secondary sexual characteristics were signalled, meant that medical discourse was once more modified and took into account what patients actually said about their bodies and desires.

Another important aspect of the relationship between patient and doctor concerns the relative importance of surgery. In the first phase of the period of the 'true sex', when it was recommended that the voice of the patient was muted and when the biological markers that proved the real sex of the person were sought, the state of surgical technology was very limited. Doctors were restricted to touching, seeing, measuring, examining orifices or exploring external genitalia, for example, the opening of the scrotal sac to detect the presence of testicles. The microscope was increasingly used to examine semen. In this way, doctors tried to 'reveal' or uncover the true sex lying below or hidden in the patient, allowing it to speak on its own terms. The patient was encouraged to allow the doctors to proceed by invoking the progress of science and the welfare of humanity as a whole.

During the second phase, from approximately 1905 to 1915, the diagnosis of hypospadias understood within a degenerationist framework at least partially supplanted that of hermaphroditism. This gave rise to multiple surgical interventions designed to reconstruct the genitals of hypospadic males. A third period, with its developments in testicular grafts, organotherapy and from the 1940s laparotomy, permitted a new perspective. True sex was identified by the gonads and the hormonal secretions were given significant influence. Pathologies such as eunuchoidism, ovarian insufficiency, cryptorchidism, various degrees of hypospadias, different types of hermaphroditism and pseudo-hermaphroditism were susceptible to treatment by surgical and pharmacological means. It was now a case of not only identifying true sex but magnifying it and eliminating those biological or chemical elements that placed it in doubt.

Finally, with the split between biological sex and psycho-social sex an immense field for another type of surgical intervention was opened up. From the second half of the 1950s the decisive question in surgical interventions no longer depended on the organism – the identification and magnification of the true sex – but on social convenience – the avoidance of the scandal of changing civil sex – and on the psychological health of the individual. It was necessary to adjust the somatic appearance of the individual to the learned (and, indeed, desired) sex of the patient. This was the period of what we have called 'sex as simulacrum'. It was not a case of revealing the hidden sex but, following the example of John Money and his team at the Johns Hopkins Hospital, to *reassign* it and eliminate the vestiges of the undesired sex, which could go on to produce psychological disturbances in the intersexed individual. From here it is a short jump to 'transsexuality' and the opening up of new gender reassignment clinics.

An important issue in many doctors' minds in several countries was the possible relationship between hermaphroditism and homosexuality and the concern harboured by effective same-sex marriages that were ascertained after examination of patients' bodies. The association between hermaphroditism and sexual transgression is a long one and can be traced back at least to the medical texts of the sixteenth and seventeenth centuries in which the 'effeminate sodomite' and the supposedly luxurious capacities of women with large genitalia are mentioned. In the early 1900s the term 'hermaphrodite' was often used popularly as a synonym for sexual invert. In the nineteenth century – in contrast to the twentieth – cases discussed in the medical press or in manuals in Spain rarely focused on the sexual preferences of the hermaphrodite subject.

However, in the twentieth century, it is common to find mention of the hermaphrodite in medical manuals in sections that refer to the sexual perversions. Except in isolated cases such as that of the case of ambiguous sex discussed by Dr Pascual in 1920, we have to wait until the 1950s for an explicit mention of the combined threat of hermaphroditism and homosexuality. At the height of Franco's rule, some intellectuals such as Mauricio Carcavilla were determined to fuse sexual dissidence with political subversion. Not even Marañón, who considered hermaphroditism and homosexuality as two variants within the intersexual paradigm and who wrote so much about the sexual question, dedicated much time to the sexual inclinations of hermaphrodites.

What does seem to have emerged during the late nineteenth and early to mid-twentieth century, in the period from the Restoration up to the end of the second Republic, is a concern about hermaphroditism as a symptom of the crumbling frontiers of the genders. The strict boundaries between male and female would have been under threat from a variety of causes including the rising feminist movement, the appearance of the new woman, the increasingly visible presence of 'sexual inverts', the crisis of regulated prostitution, the rise of the threat of venereal disease (particularly in the form of the 'devouring woman' or vamp) and the growth of a bohemian aesthetic projected as a reflection and cause of national decadence and the crisis of masculinity. Other influences included the crisis of Don Juan as the prototype of the Spanish male, the decline of the certainties of positivism (associated with the thinking, rational western man) and the rise of occultism and irrationalism (associated with the savage, eroticized oriental woman).

But this understanding of the hermaphrodite as a de-civilizing force was not uniform. The supporters of Dr Moebius in Spain such as Angel Ganivet and to some degree the pathologist Nóvoa Santos understood women as inferior beings. Stemming from this appreciation, those who advocated the gonadal criteria of hermaphroditism tended to minimize the existence of 'true hermaphrodites' in the human species, as if in this way they could dismiss the ghost of dissolution of the great sexual divide between men and women. From a slightly less old-fashioned stance liberal doctors such as Marañón were keen to bring their theories up to date, thereby preserving sexual difference as a regulatory principle in modern sex education. But this move was achieved by advancing a broad concept of 'intersexuality' developed from the zoologist Goldschmidt's work. Lack of sexual differentiation was broadened out to include gynaecomastic men, those with cryptorchidism, homosexuals, 'virile' women and androgynes, to the degree that the distinction between pseudo- and real hermaphrodites, defended to the hilt by those who supported the gonadal explanations of Klebs and Pozzi, was fundamentally undermined. A cordon sanitaire of pathological figures, blanched of their previous sinful connotations and now to be pitied for their degenerate condition, was drawn up to immunize the distinctions between the sexes, whose boundaries were not seen as impermeable.

A further matter, which enables us to place the present ethical and political problems faced by today's 'intersexuals' in Spain in historical perspective, is that of 'medical paternalism'. On paper, since the nineteenth-century legal medical attempt to fix the true sex of individuals in cases of doubtful sex, the biological sciences have stopped offering any kind of rationale – as Elena de Céspedes did in her inquisitorial trial of 1587 – that would legitimize the existence of hermaphrodites or sex changes in the human species. If medicine discovered that an individual was living as the wrong sex it was required to inform the authorities so that an official rectification could take place in the Civil Register, obliging the individual to adopt his or her 'true biological sex'. In the case of married persons, any union falsely obtained was to be annulled.

Such a perspective in the field of legal medicine, during the period when the 'true sex' reigned, permitted a frankly authoritarian form of medical intervention, which prevailed over any kind of expression of volition by the individual concerned. However, in practice, this strict following of the forensic norm (more common in France and less assiduous in Britain) does not appear to have governed Spanish doctors' actions. There were indeed cases of 'rectification' of the assigned identity, above all in children, as in the example discussed from 1908 under the auspices of Dr Morales Pérez, and in some adults. In these cases, the rectification did not appear to contrast with the understanding of self of these individuals. In child cases, a prudent period of waiting is advocated until the awakening of the erotic instinct or the time of marriage. Finally, in those few cases of marriage (for example, the one discussed by Pascual in 1920) where the sex of one partner is discovered to be the opposite of what was thought, the tendency is to leave the case open, in suspense, without necessarily pronouncing in favour of a marriage dissolution.

In the mid-1950s a different set of criteria is imposed. Rectification of sex is not favoured. From now on doctors point to the importance of the happiness or psychic well-being of the patient (a good example of this new trend is the 1960 case studied by Fontes Gil, Lorente Fernández, Jiménez Casado and Arnal Arambillet). In these cases, social convenience (the avoidance of scandal), public morality (not inciting evident homosexuality) and the dignity of the individual mean that patients are assigned the sex that they have lived with and for which they are accepted, regardless of the biological truth. It is recommended, for the individual's 'own good', that they be left in ignorance of their situation. Today we know that this practice, which guides the 'corrective plastic surgery' undertaken on infants in Spain from the 1970s to date, is not only an act of medical paternalism but also places the individual's health at risk (despite the fact that the doctor must 'inform' the parents and obtain their 'consent'). If parents are not informed or if the adult has surgical intervention even with 'informed consent', there is a risk of severe physical or psychological consequences, including the development of genital cancers.[6]

The fact is that, throughout the whole period that this book has covered, the belief that one body can only contain one sex in order to be socially and legally recognized has prevailed. This was the case during the period of 'sex as status' and it was so during the period of 'true sex', whether anatomical, gonadal, hormonal or chromosomal. That medicine recognized in the first period that hermaphrodites existed in no way undermines the fact that individuals had to choose or be assigned one sex and stick with it. Paradoxically, the truth of this belief is more evident at the present time, the period of the 'sex as simulacrum'.

In fact, the desire of the medical profession to set parents at ease by intervening surgically on intersexual babies in order to give them an acceptable appearance as a male or a female shows how far medicine will go to mask reality (the fact of intersexuality) and in order to make bodies conform to certain values (a heteronormative system of values that forces individuals to identify with one sex and one sex alone). It constitutes an attempt to respond to parents' anxieties by adjusting what is seen (the genitals) in accordance with what is said (the signifiers conventionally associated with each sex). What functions as a performative discourse and practice in the reiteration of sexual difference is raised to the level of information and pure fact.[7]

In contrast to this age-old performative gesture that reiterates in different ways the impossibility of escaping from the *dictum* of 'one body, one sex', anthropologists teach us that such a rule is far from being present in all cultures.[8] Political activists and intellectuals associated with the trans movement since the late 1980s and early 1990s question the dimorphic sexual model and speak of a rainbow of sexual differences that would give rise to a multiplicity of sexes and sexualities.[9]

Political mobilization of intersexuals began in the United States in the 1990s. Protest was organized around the surgical procedures involved in sex reassignment, including clitoridectomy practised on children. The movement demanded a change in the medical protocols governing procedures. In 1993 Cheryl Chase, who underwent clitoridectomy as a young girl in the 1960s, founded the ISNA (Intersexual Society of North America).[10] This organization has managed to change medical practices in cases of intersexuality in the States. In addition to campaigning against sex reassignment surgery in children, ISNA has fought for the depathologization of intersexuality, the struggle for the human rights of intersexuals, a critique of the binary model in conceptualizing the sexes and has rejected the label of 'gender dysphoria' for those individuals who are not in agreement with the sex the medical profession has assigned to them.[11] In 1998 the Organisation Internationale d'Intersexes (OII) was created. It was founded by Curtis Hinkle and emanated initially from the Francophone population of North America but has now expanded to include groups in twelve countries, including Spain.[12] However, in Spain, as far as we know, there is no separate organization for the defence of intersexual rights, although in some cases issues are taken on board in transsexual organizations. In October 2007 both communities mobilized to protest against the letter of the recent Gender Identity Law (in place since March 2007), which was passed under the socialist government of Rodríguez Zapatero. This law, unique in many senses in its liberal stance, does not demand change-of-sex surgery as a prerequisite for changing sex in the civil register but it does still demand that a psychiatric diagnosis takes place detailing the 'gender dysphoria' of the patient in order to change the name and sex of the person officially (with the exception of those under 18, immigrants and the mentally disabled).[13]

Despite this mobilization and an increasing number of published works on these subjects on the part of biologists such as Fausto-Sterling, historians such as Alice Dreger, sociologists such as Meira Weiss and by intersexuals like Cheryl Chase and Iain Morland, who have all contributed to the critique of sex-reassignment surgery in babies and young children, doctors continue to argue in favour of, and implement, this practice. In doing so, they continue to reinforce social prejudices with their medical protocols and continue to place the physical and mental health of their patients in question. As Iain Morland has written, intersexuality is not in itself a threat to life but rather a menace to a heteronormative culture. This particular threat does, in contrast, present real health threats to those caught up in it. In this sense, doctors' desire to operate immediately on intersex babies in order to eliminate ambiguities reflects a similar aspiration to that of their nineteenth-century colleagues who were determined to locate the 'true' genitals and 'true' gonads in their patients.

Proof of the persistence of these kinds of prejudice and of the harm they may cause (including anorgasmia as a result of clitoral reduction or extirpation and cancer) is shown by a recent case of 'hermaphroditism' in a new-born in Cadiz. On 4 April 2007 a baby of indeterminate sex was born to working-class parents from the Santa María quarter of the city of Cadiz. Doctors at the University Hospital of Puerta del Mar were quickly on hand to deal with the situation. As a local newspaper reported: 'specialists from the Neo-Natal Department ... are undertaking various chromosomal tests in order to proceed to an operation so that the baby will grow up with the type of sex that predominates'.[14] 'True sex' is not referred to but 'predominant sex' is and it is hoped, according to the journalist, that the genetic study will resolved the enigma. It is acknowledged, however, that one variety of intersexuality is present in those individuals who are genetically (XY) and in terms of gonads are male but who, as a result of androgen insensitivity, take on a generally feminine aspect; it is this sex that is usually assigned to them.

Although the case is still under urgent review, the journalist notes that the operation will take place so that the baby grows up like a man, even though it is still necessary to wait for the results of the studies as, in order to come to a decision, other factors beyond the external genitalia must be taken into account. In fact, the specialists believe that the chromosomal, neural, hormonal, psychological and behavioural aspects are more important.

The situation is essentially paradoxical – in order to satisfy the parents and society surgical intervention is rapidly announced but, at the same time, the fragility of such a decision is alluded to. How is it possible to take into account the psychological and behavioural traits of a baby? Is not the decision already taken that the baby will be brought up as a male, given the appearance that the genitalia will be given?

A second news item on this case allows for a fuller account. Two days after the birth, the same journalist from *La Voz de Cádiz* published an interview with a psychologist who had offered free advice to the parents of the hermaphrodite baby. The psychologist, the director of the Institute of Sexology in Seville, identified the two options available in order to determine the sex of the baby. The first option, 'the most common, is taken during the first few months of life'.[15] This involved surgical intervention to give the baby the appearance of one sex or the other. This was a procedure that was more common in those cases accepted as female. The testicles would be extirpated, genitoplasty, which may include the reduction of the size of the clitoris as in the case of Cheryl Chase, would take place and treatment with oestrogen would continue throughout life.

The second option is preferred by the psychologist. It entailed waiting somewhat longer and 'the assignation of sex would take place as gender identity developed, thus allowing children to grow up in an intersex condition postponing the decision until a time when they were old enough and in accordance with the male or female gender identity' that emerged.[16] It was admitted, however, that 'there are legislative problems in this country that impede this particular option'. This is to say that the first option is adopted not for medical reasons but for legal reasons. The existing legal framework confirms the old prejudice that people cannot exist, even for a short period of some years, with two sexes at the same time. The medical staff have no option but to reiterate this as they prepare hurriedly to intervene surgically. The psychologist recognizes that 'there exist many cases in which one sex has been assigned at birth and in whom during adulthood they present discordances with their gender identity'. With this statement, the psychologist puts his finger on the defining point of the Gender Identity Law: this law does not alter the fact that any person who wishes to change their sex in adulthood must be diagnosed with gender dysphoria in order to be able to do so.[17]

Finally, on the subject of the benefits of surgery in order to confirm one sex or the other, the psychologist essentially reaffirms the categories used by Klebs ('male pseudo-hermaphroditism', 'female pseudo-hermaphroditism' and 'true hermaphroditism'), even though they are presented with different vocabulary (Morris, Reifenstein, Swyer, Turner and Klinefelter syndromes). In the twenty-first century the conceptual frameworks employed in practical diagnostic terms have much in common with those of the Age of Gonads, that is, the notion of real or true sex still applies despite its contradiction of the importance of 'psycho-social sex' and despite the ultimately contradictory terms such as pseudo-hermaphroditism which are destroyed in the act of their own enunciation.[18]

The analysis offered in this book allows us to reflect from a historical perspective on the legacies and strains of thought from the past that continue to inform present-day expert discourse on intersexuality and hermaphroditism. Such a historical overview enables us to see how contingent such diagnostics and practices are, despite their aspiration to objectivity and truth.

As we have stated often throughout this book, the objective of this piece of research was not to apply letter by letter to the Spanish case those historical models developed for other countries such as Britain and France. The otherwise excellent account of Alice Dreger, for example, is overschematic in our view about the stages of hermaphrodite science as it proceeds from the period of the true anatomical sex through to that of the truth of the gonads. In the Spanish case, it can be said with confidence that these stages were passed through much less definitively and with other characteristics. Many different ways of seeing and doing were recorded. In what remains of this final chapter, it is worth summarizing the differences between the Spanish case and those of especially Britain and France in respect of the science of hermaphroditism. Principally, there are, in our view, five major differences worth pointing to.

First, in Spanish medicine from the sixteenth and seventeenth centuries there predominated a variant in the Hippocratic-Galenic representation of the 'one-sex' model. Some examples have been recorded which modify or openly critique this unitary model, both in the literatures of magic and marvels (Martín del Río, Fuentelapeña) and in anatomical treatises (Luis Mercado, García Carrero, Bravo de Sobremonte). What are not to be found in Spain are traces of a dualist clearly Aristotelian model such as that defended in the sixteenth century by the Italian anatomist Benedetto Varchi or by the French doctor Jean Riolan in the seventeenth.

Secondly, while in eighteenth-century medicine and naturalist thought, which favoured the 'expulsion of the marvellous', there is an evident refusal to accept the existence of sex changes (Barco y Gasca), the refusal to accept actual hermaphrodites is more exceptional (Hervás y Pandero, Fernández del Valle). In Spain, the more radical refusal to accept hermaphrodites that was adopted by other European thinkers of the Enlightenment (see, for example, the article on hermaphroditism by James Parsons in the *Encyclopédie*) was not present.

Thirdly, in the nineteenth century with the 'true sex' period in full swing, Spanish doctors seem to be less keen to expose 'errors of sex' than their French counterparts. This more cautious attitude of the Spanish would place them alongside the British doctors and they were more unwilling to alter the sex, particularly of adult persons, with full legal effect in the name of biological truth. The same degree of caution is encountered in Spanish doctors' approach to sanctioning surgical intervention to rectify sex. The case presented by Dr Morales in 1908 is paradigmatic of this stance. The sex of an 8-year-old boy was corrected by changing the name, clothing and activity of the child but no surgical intervention was advocated in order to change the appearance of the genitalia.

Fourth, at least up to the middle of the twentieth century, Spanish doctors were more reticent in their questioning of subjects about sexual desires and practices. This again was in contrast with their British and French colleagues. A similar late use of photographs is also to be recorded; their generalized use came in the 1940s and 1950s in Spain.

Fifth, with respect to the reception of the taxonomy drawn up by Klebs, the concept of pseudo-hermaphroditism and the primacy of gonadal sex, Spain presents some peculiarities. On the one hand, the use of Klebs's concepts seems to have occurred earlier than in France or Britain. However, the norm of gonadal sex appears to have been consolidated later than in the two aforementioned countries. It was never a hegemonic concept in Spain and had to compete with other concepts such as hypospadias, the hormonal criteria and Marañón's theory of intersexuality. Finally, the term pseudo-hermaphroditism, which gradually fell into disuse in France and Britain in the 1920s, was curiously revitalized in Spain in the latter part of the same decade.

These summary comparative conclusions, which could be expanded if there were more studies on hermaphroditism from other latitudes and cultures,[19] allow us to counter any idea that, in the Spanish case, the science of hermaphroditism was behind the times relative to other European countries. It is true to say that the Spanish scientific community did not excel in creating new conceptual frameworks for the understanding of hermaphroditism, with the obvious exception, to some degree, of Marañón's hybrid theory. The criteria drawn up by Marc, the taxonomies of Geoffroy de St-Hilaire, the embryological theories of Meckel, the typologies created by Klebs and Pozzi, the innovations of endocrinological science and the work of John Money and his team, all important steps in the development of this field, suffered no particular delay in reception amongst Spanish medical figures. Despite the terrible impact of the Civil War and the arrival of Francoism, the lack of innovation and the more peripheral character of the Spanish contribution to the field are not sufficient to support the myth of Spanish backwardness in scientific or sexological terms. The Catholic Church, despite what one might expect its role to be, is not particularly evident or present. In comparison to its role in other areas connected to sexuality – the debate on 'free love', the campaigns on sex education, the crusade against neo-Malthusianism and pornography – the hermaphrodite question in Spain does not appear to have turned the ecclesiastical head. The question of hermaphroditism was, no doubt, a delicate subject area where all discretion was called for and where any realization of error or change of sex would have been met with a veil of silence. At the same time, this subject was understood to be a specialized area, an ambit with its own problems and questions, reserved to the more technical language and more inaccessible techniques of trained professionals. It was sufficient to recall that God had created two sexes in the human species and that any alteration in that order was nothing but a mistake of minor importance.

Caught up in the snares of their expert custodians, who took on the mission of guaranteeing that all Spaniards possessed one sex and one sex alone, hermaphrodites in Spain became the plaything of triumphant science and served to test out reassignment surgeries and psycho-sexual models of functionality. Little by little, however, and often accompanied by politicized lesbians, gays and transsexuals, intersexuals have made themselves heard. One of the aims of this book is to accompany and amplify that voice.

Notes

1 Here we reject the supposed dichotomy between pure science and applied science and consider scientific activity as a complex practice inserted into *dispositifs* or *assemblages* of heterogeneous elements that produce forms of objectivity (in this case, the 'hermaphrodite'). On the pure/applied science question, see Latour (1987) and Echeverría (1994). On the 'pragmatic turn' in epistemology, see Rehg and Bohman (2001) and Schatzki et al. (2001). On *assemblages*, see Rose (1999: p. xv) and Dean (1999: 29–30). On the Foucauldian roots of the *dispositifs* or 'social apparatuses' see Deleuze (1992).
2 The 'iceberg' and 'kaleidoscope' images come from Veyne (1984: 203–22).
3 On the resignification of gender in this sense, see Butler (1993: 122–4).
4 Foucault (1990b: 175–202).
5 Boltanski (1975: 59–66). This is one aspect absent from the excellent work of Alice D. Dreger – the lack of attention paid towards bodily cultures and the effect of class differences on these.
6 On these questions, see Fausto-Sterling (2000: 79–95); Dreger (1998: 167–201); Chase (1998).
7 On this point, see the suggestive article by Morland (2001: 529–30).
8 On the cases of Third Sex in the Sambia in New Guinea, the Polynesian Mahu, the Indian hijras, the Zuni and others, see Herdt (1996); Roscoe (1991); Nieto (1998, 2004).
9 Fausto-Sterling (2000: 95–114). Other authors in this vein are Cheryl Chase, John Colapinto, Milton Diamond, Alice D. Dreger, Morgan Holmes, Suzanne Kessler and Iain Morland.
10 See Fausto-Sterling (2000: 180–4); Dreger (1998: 176–8).
11 On these demands see the Argentine page of the intersex organization *http://www.intersexualite.org/Spanish-Index.html*.
12 *http://www.intersexualite.org*.
13 A recent article discussing the positive aspects of this law and some of its negative consequences is Platero (2008).
14 Agrafojo (2007a).
15 Agrafojo (2007b).
16 In the Argentinean/Spanish/French film *XXY* (dir. Lucía Puente,

2007), the protagonist, Alex, effectively chooses not to choose between the sexes and to live with the consequences with the support of the father, played by Ricardo Darín. Despite what we say about the Cadiz case, we do not wish to underestimate the very difficult situation faced by the parents of the baby born to them in 2007. Rather, we question the medical protocol and prerogatives, and all the pressure brought to bear, that cases such as this entail.

17 Platero (2008: 47).
18 On the contradictory and androcentric nature of these categories, see Morland (2001: 534–9).
19 In Italy, from 1900 until the end of fascism, it is possible that there existed an interest in hermaphroditism similar to that in Spain. The work by Weininger, *Sex and Character*, translated into Italian in 1912, was republished five times over the two decades of the fascist regime. There was less a fear of homosexuality than of an erosion of the borders between the two sexes. On this question, see Wanrooij (1990: 208–10).

Bibliography

Spellings and authors' names, sometimes lacking initials, are as in the original. The following abbreviations are used:

AMCE	*Archivos de Medicina, Cirugía y Especialidades*
EM	*La España Médica*
MI	*La Medicina Ibera*
RMCP	*Revista de Medicina y Cirugía Prácticas*
SM	*El Siglo Médico*

Ackerknecht, E. H. (1967) *Medicine at the Paris Hospital* (Baltimore, MD: Johns Hopkins University Press).
Agamben, G. (1998) *Homo Sacer: Sovereign Power and Bare Life*, trans. Daniel Heller-Roazen (Stanford: Stanford University Press).
Agrafojo, N. (2007a) 'Nace en Cádiz un bebé hermafrodita y los médicos estudian qué sexo predomina en él', *La Voz de Cádiz* (22 April 2007).
Agrafojo, N. (2007b) 'José Luis Sánchez de Cueto Lorenzo, Psicólogo y Sexólogo: "Por encima de todo esto debe estar el bienestar del ser humano"', *La Voz de Cádiz* (24 April 2004).
Alba y López, R. (1860) 'Caso de hermafrodismo presentado a la consulta clínica del Dr. Ulibarri', *EM*, 5, 265.
Alba y López, R. (1861) 'Operación practicado por el Dr. Ulibarri en el caso de hermafrodismo que ha habido en su clínica. Reflexiones acerca de ella', *EM*, 6, 455–6.
Alberola, M. (1988) 'L'hermafrodita guerriller', *El Temps* (29 Feb.–5 March) available at *http://memoriacastello.cat/pagina1.22.html* (consulted Jan. 2008).
Alberti López, L. (1948) *La anatomía y los anatomistas españoles del Renacimiento* (Madrid. CSIC).
Albucasis (1973) *On Surgery and Instruments*, ed. M. S. Spink and G. L. Lewis (London: Wellcome Institute of the History of Medicine).
Aldaraca, B. (1991) *El ángel del hogar: Galdós and the Ideology of Domesticity in Spain* (Chapel Hill, NC, and Valencia: University of North Carolina, Dept of Romance Languages, and Artes Gráficas Soler).
Alemán, M. (1968) *Guzmán de Alfarache*, in *La Novela Picaresca Española*, ed. A. Valbuena y Prat (Madrid: Aguilar), pp. 159–637.
Alfonso X El Sabio (1843–4) *Las Siete Partidas ... con las variantes de más interés y con la glosa del Licenciado Gregorio López*, vol. 4, partida VI, tit. 1, ley 10, ed. Ignacio Samponts (Barcelona: Ramón Martí y J. Ferrer).
Alvarez de Miravall, B. (1597) *Libro intitulado La Conservación de la Salud del Cuerpo y del Alma* (Medina del Campo: S. del Canto).

Androutsos, G. (2003) 'Louis Ombrédanne (1871–1956) et la cure de l'hypospadias', *Progrès en Urologie*, 13, 277–84.
Anon. (1841) 'Medicina legal: nueva aplicacion del microscopio á los esperimentos médico-legales', *Boletín de Medicina, Cirujía y Farmacia*, 2nd ser., 66, 237.
Anon. (1842) 'Reglamento aprobado por S. M. el Regente del Reino para la declaracion de exenciones físicas del servicio militar', *Boletín de Medicina, Cirujía y Farmacia*, 2nd ser., 99, 210–13.
Anon. (1861) 'Hipospadias: observaciones acerca de algunas variedades de esta enfermedad y del tratamiento quirúrjico que les conviene', *SM*, 370(8), 74.
Anon. (1863) 'Bibliografía', *EM*, 7, 547–8.
Anon. (1865) '¿El acto fecundante es el acto en que se determina el sexo?', *EM*, 9, 327.
Anon. (1879) 'Tratamiento del hipospadias y del epispadias', *RMCP*, 5, 520–1.
Anon. (1880) [No title], *Gaceta Médica de Sevilla* (4 April), 122–3.
Anon. (1881) 'Hermafrodismo ó hypospadias', *RMCP*, 7, 181–2.
Anon. (1888) 'Un caso de hermafrodismo aparente', *RMCP*, 22, 312–13.
Anon. (1890) 'Pseudo-hermafrodismo masculino', *RMCP*, 26, 273–4.
Anon. (1891) 'Un caso de hermafrodismo', *RMCP*, 28, 645–6.
Anon. (1906) 'Patología y tratamiento del criptorquidismo', *RMCP*, 70, 83.
Anon. (1907) 'Hipospadias peno-escrotal', *RMCP*, 74, 122.
Anon. (1909) 'Hipospodia [sic] acompañado de otros vicios de desarrollo', *RMCP*, 82, 205.
Anon. (1910a) 'Hipospadias peniano. Autoplastia uretral', *RMCP*, 88, 279.
Anon. (1910b) 'Hipospadias peno-escrotal', *RMCP*, 88, 284–5.
Anon. (1910c) 'Resultados de los distintos métodos operatorios del hipospadias', *RMCP*, 88, 117–18.
Anon. (1910d) 'Hermafroditismo', *RMCP*, 86, 1910, 158.
Anon. (1919) 'Un caso de hypospadias completo', *SM*, 3421, 542–3.
Anon. (1920a) 'Un caso de hermafrodismo', *MI*, 11 (135), 181.
Anon. (1920b) 'Un caso humano de hermafrodismo bilateral de glándulas bisexuales', *MI*, 12(139), 264.
Anon. (1923) 'Hermafroditismo (verdadero) glandular en un individuo de diez años', *MI*, 17(293), 550.
Anon. (1925) Review of Salvador Pascual, 'Los hermafroditas', *Archivos de Endocrinología y Nutrición*, 1, 1925, in *MI*, 19(395), 564–5.
Anon. (1931) 'Sobre un caso de seudohermafroditismo', in 'Sesiones Científicas: Academia Nacional de Medicina. 21 de marzo de 1931', *AMCE*, 34(20), 464–5.
Aresti, N. (2001) *Médicos, donjuanes y mujeres modernas: Los ideales de feminidad y masculinidad en el primer tercio del siglo XX* (Bilbao: Universidad del País Vasco/Euskal Herriko Unibertsitatea).
Aresti Esteban, N. (2007) 'La mujer moderna, el tercer sexo y la bohemia en los años veinte', in Rosa Monlleó and Jordi Luengo (eds), 'Espais de Bohèmia: Actrius, Cupletistes i ballarines', *Dossiers Feministes*, 10, 173–85.

Aristotle (1943) *Generation of Animals*, book 4, trans. A. L. Peck (London and Cambridge, MA: William Heinemann and Harvard University Press).
Ashcom, B. B. (1960) 'Concerning "la mujer en hábito de hombre" in the Comedia', *Hispanic Review*, 28, 43–62.
Bakhtin, M. (1987) *La cultura popular en la Edad Media y en el Renacimiento* (Madrid: Alianza Universidad).
Balthazard, V. (1947) *Manual de medicina legal* (Barcelona: Salvat; 1st publ. 1906).
Baldwin, J. W. (1994) *The Language of Sex: Five Voices from Northern France around 1200* (Chicago: University of Chicago Press).
Bañuelos, M. (1941) *Manual de patología médica*, 2 vols (Madrid: Editorial Científico Médica).
Barbazza, M. C. (1984) 'Un caso de subversión social: el proceso de Helena de Céspedes (1587–1589)', *Criticón*, 26, 17–40.
Barnes, B., D. Bloor and J. Henry (1996) *Scientific Knowledge: A Sociological Analysis* (London: Athlone).
Bauer, J. (1930) *Herencia y constitución* (Barcelona: Labor).
Bayo, C. (1902) *Higiene sexual del soltero* (Madrid: Librería de Antonio Rubiños).
Beusterien, J. L. (1999) 'Jewish male menstruation in seventeenth-century Spain', *Bulletin of the History of Medicine*, 73(3), 447–56.
Birke, L. (1999) *Feminism and the Biological Body* (Edinburgh: Edinburgh University Press).
Bloom, A., and S. Benardete (eds) (2001) *Plato's 'Symposium'*, trans. Seth Benardete (Chicago and London: University of Chicago Press).
Boltanski, L. (1975) *Los usos sociales del cuerpo* (Buenos Aires: Ediciones Periferia).
Borell, M. (1976) 'Organotherapy, British physiology, and discovery of the internal secretions', *Journal of the History of Biology*, 9(2), 235–68.
Borell, M. (1985) 'Organotherapy and the emergence of reproductive endocrinology', *Journal of the History of Biology*, 18(1), 1–30.
Boswell, J. (1980), *Christianity, Social Tolerance and Homosexuality: Gay People in Western Europe from the Beginning of the Christian Era to the Fourteenth Century* (Chicago and London: Chicago University Press).
Botella Llusiá, J. (1941) 'Correlaciones endocrinas interrenal-genitales', *Revista Clínica Española*, 3 (4), 309–13.
Botella Llusiá, J. (1943) 'Problemas actuales de la maternología en la obra nacionalsindicalista de protección a la madre y el niño', *Ser*, 53–60.
Botella Llusiá, J. (1944) 'Los problemas demográficos de la maternología española', *Ser*, 47–52.
Botella Llusiá, J. (1946) 'El virilismo suprarrenal y síndromes afines', *Actualidad Médica*, 32(202), 604.
Botella Llusiá, J. (1953) 'Los hermafroditas: una contribución a la cirugía de sexo', *Revista de la Universidad de Madrid*, 1 (3), 363–73.
Botella Llusiá, J. (1955) *Fisiología Femenina*, 3rd edn (Barcelona and Madrid: Editorial Científico Técnica).
Botella Lluisá [sic], J. (1956) *Dos casos de hermafroditismo verdadero: Conferencia*

en la Real Academia Nacional de Medicina. *Día 24 de enero de 1956* (Madrid: Imprenta de José Luis Cosano).
Botella Llusiá, J. (1966) 'El hermafroditismo verdadero', *Acta Ginecológica*, 17(5), 279–98.
Botella Llusiá, J. (1971) '¿Existe una cibernética de la reproducción humana?', *Revista de la Universidad de Madrid*, 20(77), 71–80.
Botella Llusiá, J. (1983) 'Sociobiología del sexo', *Arbor*, 14(446), 7–19.
Botella Llusiá, J. (1987) 'Prólogo', in J. Martínez Castellanos, *Intersexos: estados intersexuales* (Madrid: Gráficas Sebastián).
Botella Llusiá, J. and F. Nogales (1952) 'Sobre el síndrome de seudohermafroditismo masculino con feminización total', *Acta Ginecológica*, 3, 319.
Bourdieu, P. (1982) *Leçon sur la leçon* (Paris : Minuit).
Bouza, F. (1991) *Locos, enanos y hombres de placer en la Corte de los Austrias* (Madrid: Temas de Hoy).
Boylan, M. (1984) 'The Galenic and Hippocratic challenges to Aristotle's conception theory', *Journal of the History of Biology*, 17(1), 83–112.
Brachfeld, F. O. (1932) 'Crítica de las teorías sexuales del Dr. Marañón', *SM*, 4081, 214–21.
Bravo de Sobremonte, G. (1671) 'Promptuarium XXIV', 'De sexus mutatione' and 'De hermaphroditis', in *Operum Medicinalium*, vol. 3 (Lyon: L. Arnaud), pp. 246–9 and 249–50.
Bravo de Sobremonte, G. (1679) 'Resolutio I. Utrum sexus transmutatio permitatur naturae', in *Operum Medicinalium*, vol. 4, *Tres Disputationes Complectens* (Lyon: L. Arnaud), p. 198.
Bravo Villasante, C. (1955) *La mujer vestida de hombre en el teatro español (siglos XVI–XVII)* (Madrid: Revista de Occidente).
Briand, J., J. Bouis and J. L. Casper (1872–3) *Manual completo de medicina legal y toxicología*, vol. 2 (Madrid: R. Labajos).
Brisson, L. (2002) *Sexual Ambivalence: Androgyny and Hermaphroditism in Graeco-Roman Antiquity* (Berkeley–Los Angeles and London: University of California Press).
Brown, P. (1988) *The Body and Society: Men, Women and Sexual Renunciation in Early Christianity* (New York: Columbia University Press).
Burshatin, I. (1999) 'Written on the body: slave or hermaphrodite in sixteenth-century Spain', in J. Blackmore and G. S. Hutcheson (eds), *Queer Iberia: Sexualities, Cultures and Crossings from the Middle Ages to the Renaissance* (Durham, NC: Duke University Press), 420–56.
Butler, J. (1993) *Bodies that Matter: On the Discursive Limits of 'Sex'* (New York and London: Routledge).
Butler, J. (2004) *Undoing Gender* (New York and London: Routledge).
Cacho Viu, V. (1997) *Repensar el 98* (Madrid: Biblioteca Nueva).
Cadden, J. (1993) *Meanings of Sex Difference in the Middle Ages: Medicine, Science, and Culture* (Cambridge: Cambridge University Press).
Camacho Alejandre, F. (1913) 'Un caso de seudohermafrodismo en un niño hipospádico y criptórquido', *La Actualidad Médica* (Granada), 265–70.
Campos Marín, R. (1999a) 'La teoría de la degeneración y la clínica psiquiátrica en la España de la Restauración', *Dynamis*, 19, 429–56.

Campos Marín, R. (1999b) 'La teoría de la degeneración y la profesionalización de la psiquiatría en España (1876–1920)', *Asclepio*, 51(1), 185–203.
Cañadell, J. M., and J. Planas Guasch (1961) 'Hermafroditismo verdadero', *Medicina Clínica*, 37(5), 335–44.
Canguilhem, G. (1980) 'La Monstruosité et le monstrueux', in *La Connaissance de la Vie* (Paris: Vrin), pp. 178–9.
Cardenal, L. (1924) 'Un caso de pseudohermafroditismo', *MI*, 18(339), 439.
Carlavilla del Barrio, M. (1956) *Sodomitas* (Madrid: Editorial Nos).
Carranza, A. (1630) *De Partu Naturali et Legitimo* (Jacobi Stor: Cologne).
Carter, J. (1997) 'Normality, whiteness, authorship: evolutionary sexology and the primitive pervert', in V. A. Rosario (ed.), *The Erotic Imagination: French Histories of Perversity* (New York and Oxford: Oxford University Press), pp. 155–76.
Casper, J. L. (1886–7) *Tratado práctico de medicina legal*, trans. F. Alvarez-Ossorio y Pizarro, abogado of the Colegio de Madrid, 5 vols (Madrid: Est. Tip. de Pedro Nuñez).
Castillo Martín, M. (2006) 'Escritoras y periodistas en los años veinte', in I. Morant (ed.), *Historia de las Mujeres en España y América Latina*, vol. 4, *Del siglo XX a los umbrales del XXI* (Madrid: Cátedra), pp. 169–90.
Castro Pires de Lima, A. (1958) *A Mulher vestida de Homem (Contribuição para o estudo do romance 'A Donzela que vi à Guerra')* (Coimbra: Fundação Nacional para a Alegria no Trabalho-Gabinete de Etnografia).
Cátedra, M. (2001) 'Sobre la ambigüedad: el caso de Paula Barbada', in E. Crespo and C. Soldevilla (eds), *La constitución social de la subjetividad* (Madrid: Los Libros de La Catarata), pp. 131–44.
Cayetano del Toro y Quartiellers, P. (1876) *Programa de un curso teórico-práctico de obstetricia, y enfermedades de las mujeres y de los niños*, 2 vols (Cadiz: Tip. la Mercantil).
Charnon-Deutsch, L. (1996) 'Ficciones de lo femenino en la prensa española del fin del siglo XIX', in Iris M. Zavala (ed.), *Breve historia feminista de la literatura española (en lengua castellana)*, vol. 3, *La mujer en la literatura española: Modos de representación desde el siglo XVIII a la actualidad* (Barcelona and San Juan: Anthropos and Universidad de Puerto Rico), pp. 49–79.
Chase, C. (1998) 'Hermaphrodites with attitude: mapping the emergence of intersex political activism', *GLQ: A Journal of Gay and Lesbian Studies*, 4(2), 189–211.
Chinchilla, A. (1967) *Anales históricos de la medicina en general y biográficos-bibliográficos de la española en particular: Historia de la Medicina Española*, vol. 2 (New York and London: Johnson Reprint Corporation).
Cifuentes Delatte, L. (1941) 'Sobre las hormonas sexuales masculinas', *Revista Clínica Española*, 2(1), 1–18.
Claret Miranda, J. (2006) *El Atroz Desmoche: La destrucción de la Universidad Española por el Franquismo 1936–1945* (Barcelona: Crítica).
Clavero Núñez, A. (1942) *Esterilidad Matrimonial* (Barcelona: Salvat).
Cleminson, R., and R. M. Medina Doménech (2004) '¿Mujer u hombre? Hermafroditismo, tecnologías médicas e identificación del sexo en España, 1860–1925', *Dynamis*, 24, 53–91.

Cleminson, R., and F. Vázquez García (2007) *'Los Invisibles': A History of Male Homosexuality in Spain, 1850–1939* (Cardiff: University of Wales Press).
Codina Castellví (1907) 'Pseudohermafrodismo', *RMCP*, 74, 196.
Codina Castellví (1909) 'Pseudohermafrodismo de apariencia masculina: sarcoma de ovario. Ablación', *RMCP*, 83, 282.
Codina Castellví (1932) 'La asimetría mamaria y el neumotórax', *Archivos de Medicina, Cirugía y Especialidades*, 33(11), 216–17.
Comenge, L. (n.d. [*c.*1913]) *Generación y crianza ó higiene de la familia* (Barcelona: José Espasa, Editor).
Comte Masía, M. ed. (2000) *Libro de oro de coplas y romances de ciego* (Madrid: Ediciones Añil).
Cónill Montobbio, V. (1946) *Tratado de ginecología y de técnica terapéutica ginecológica* (Barcelona: Labor).
Connell, R. W. (2001) 'Bodies, intellectuals and world society', in N. Watson and S. Cunningham-Burley (eds), *Reframing the Body* (Houndmills: Palgrave), pp. 13–28.
Cordoba, P. (1987) 'L'homme enceint de Grenade: contribution à un dossier d'histoire culturelle', *Mélanges de la Casa de Velázquez*, 23, 307–30.
Correa Calderón, E. (1961) *Baltasar Gracián: Si vida y su obra* (Madrid: Gredos).
Criado y Aguilar, F. (1932a) 'Refutación de las teorías de la intersexualidad', *SM*, 4085, 321–5.
Criado y Aguilar, F. (1932b) 'Refutación de las teorías de la intersexualidad', *SM*, 4090, 457–63.
Covarrubias, S. (1979 [1611]) 'Ermaphrodito', in *Tesoro de la lengua castellana o española* (Turner: Madrid).
Darmon, P. (1985) *Trial by Impotence* (London: Chatto & Windus).
Daston, L. (1991) 'Marvelous facts and miraculous evidence in early modern Europe', *Critical Inquiry*, 18, 93–124.
Daston, L., and K. Park (1985) 'Hermaphrodites in Renaissance France', *Critical Matrix: Princeton Working Papers in Women's Studies*, 1(5), 1–19.
Daston, L., and K. Park (1995) 'The hermaphrodite and the orders of nature: sexual ambiguity in early modern France', *GLQ: A Journal of Gay and Lesbian Studies*, 1(4), 420–5.
Daston, L., and K. Park (1996) 'The hermaphrodite and the orders of nature: sexual ambiguity in early modern France', in L. Fradenburg and C. Freccero (eds), *Premodern Sexualities* (New York and London: Routledge), pp. 117–36.
Daston, L., and K. Park (1998) *Wonders and the Order of Nature, 1150–1750* (New York: Zone Books).
Davidson, A. I. (1987) 'Sex and the emergence of sexuality', *Critical Inquiry*, 14(1), 16–48.
Davis, N. Z. (1984) *Il ritorno di Martin Guerre: un caso di doppia identità nella Francia del Cinquecento* (Turin: Einaudi).
De Arreaga, J. (1746) *Piscator Murciano: Con un agregado de prodigios, cosas no comunes y fuera de el estado natural, que han sucedido, y dignas de que se sepan, como haberse buelto muchas mujeres hombres* (Madrid: n.publ.).

De Centellas, L. (1987 [*c*.1552]) *Coplas de Don Luis de Centellas sobre la Piedra Filosofal*, in *Cinco tratados españoles de Alquimia* (Madrid: Tecnos).
De Fuentelapeña, A. (1978 [1676]) *El ente dilucidado: tratado de monstruos y fantasmas* (Madrid: Editora Nacional).
De Fuentes, A. (1547) *Summa de philosophía natural* (Seville: J. León).
De Granada, L. (1989 [1583]) *Introducción al símbolo de la fe* (Madrid: Cátedra).
De la Cerda, J. (1599) *Libro intitulado vida política de todos los estados de mujeres* (Alcalá de Henares: Juan Gracián).
De la Flor, F. (1999) 'La "puella pilosa": representaciones de la alteridad femenina (de Sánchez Cotán a José de Ribera, pasando por Sebastián de Covarrubias)', in *La península metafísica: arte, literatura y pensamiento en la España de la Contrarreforma* (Madrid: Biblioteca Nueva), pp. 267–305.
De la Pascua Sánchez, M. J. (2004) '¿Hombres vueltos al revés? Una historia sobre la construcción de la identidad sexual en el siglo XVIII', in M. J. de la Pascua Sánchez (ed.), *Mujer y deseo* (Cadiz: Universidad de Cádiz), pp. 431–44.
De León, A. (1590) *Libro primero de anatomía: recopilaciones y examen general* (Baeza: Juan Baptista de Montoya).
De Letamendi, J. (1894) *Curso de clínica general*, 2 vols (Madrid: Imp. de los Sucesores de Cuesta).
De Pineda, J. (1589) *Treinta y cinco diálogos familiares de agricultura cristiana* (Salamanca: Pedro de Adurça y Diego López).
De Torquemada, A. (1943 [1570]) *Jardín de flores curiosas* (Madrid: Sociedad de Bibliófilos Españoles).
De Torreblanca y Villalpando, F. (1678) *Epithomes delictorum sive de Magia* (Lyon: Juan Antonio Huguet).
Dean, M. (1994) *Critical and Effective Histories: Foucault's Methods and Historical Sociology* (London and New York: Routledge).
Dean, M. (1999) *Governmentality: Power and Rule in Modern Society* (London: Sage).
Dean-Jones, L. (1994) *Women's Bodies and Classical Greek Science* (Oxford: Clarendon Press).
Del Castillo Ruiz, R. (1910) 'Revista de ginecología y cirugía abdominal', *RMCP*, 86, 219–23.
Del Valle, R. (1909) 'Pseudohermafroditismo femenino externo', *RMCP*, 84, 326.
Del Valle, R. (1910) 'Un caso de sexo dudoso', *RMCP*, 86, 41.
Delcourt, M. (1970) *Hermafrodita: Mitos y ritos de la sexualidad en la Antigüedad Clásica* (Barcelona: Seix Barral).
Deleuze, G. (1992) 'What is a dispositif?', in T. J. Armstrong (ed. and trans.), *Michel Foucault: Philosopher* (Hemel Hempstead: Harvester Wheatsheaf), pp. 159–66.
Di Febo, G. (1976) 'Orígenes del debate feminista en España: la escuela krausista y la Institución Libre de Enseñanza (1870–1890)', *Sistema*, 12, 49–82.
Di Febo, G. (2003) '"Nuevo Estado", nacionalcatolicismo y género', in G. Nielfa Cristóbal (ed.), *Mujeres y hombres en la España franquista:*

sociedad, economía, política, cultura (Madrid: Editorial Complutense), pp. 19–44.
Di Febo, G. (2006) '"La cuna, la cruz y la bandera": primer franquismo y modelos de género', in I. Morant (ed.), *Historia de las mujeres en España y América Latina*, vol. 4, *Del siglo XX a los umbrales del XXI* (Madrid: Cátedra), pp. 217–18.
Díaz, E. (1974) *Notas para una historia del pensamiento español actual (1933–1973)* (Madrid: Edicusa).
Díaz del Villar, M. (1943) 'Secreciones internas de los testículos y próstata', *Gaceta Médica Española*, 8, 25–7.
Díaz Gito, M. A. (1990) 'El poema "Corsica" de J. Cristóbal Calvete de Estrella (y otros dos poemas latinos)', unpublished doctoral thesis (Cadiz: University of Cadiz).
Dreger, A. D. (1997) 'Hermaphrodites in love: the truth of the gonads', in Vernon A. Rosario (ed.), *The Erotic Imagination: French Histories of Perversity* (New York and Oxford: Oxford University Press), pp. 46–66.
Dreger, A. D. (1998) *Hermaphrodites and the Medical Invention of Sex* (Cambridge, MA, and London: Harvard University Press).
Echeverría, J. (1994) *Telépolis* (Barcelona: Destino).
Editorial (1955) 'La determinación del sexo cromosómico en la clínica', *Revista Clínica Española*, 59(4), 270.
Eiximenis, F. de (1542) *Carro de las Donas, trata de la vida y muerte del hombre cristiano* (Valladolid: Juan de Villaquirán).
Eliade, M. (1984) *Mefistófeles y el andrógino* (Barcelona: Labor).
Elias, N. (1983) *The Court Society*, trans. Edmund Jephcott (Oxford: Blackwell).
Engle (1938) 'Hormonas gonadotropas del lóbulo anterior de la hipófisis en la sangre y en la orina', *Actualidad Médica*, 24(161), 197–8.
Epstein, J. (1990) 'Either/or-neither/both: sexual ambiguity and the ideology of gender', *Genders*, 7, 99–142.
Escalonilla López, R. A. (1998) 'Función del travestismo en las comedias de Don Pedro Calderón de la Barca', unpublished doctoral thesis (Madrid: Universidad Politécnica de Madrid).
Escamilla, M. (1985) 'A propos d'un dossier inquisitorial des environs de 1590: les étranges amours d'un hermaphrodite', in A. Redondo (ed.), *Amours légitimes, amours illégitimes en Espagne (XVIe–XVIIe siècles)* (Paris: Pub. de la Sorbonne), pp. 167–82.
Eslava Galán, J. (1987) 'Introducción general', in *Cinco tratados españoles de Alquimia* (Madrid: Tecnos), pp. 13–47.
Eslava Galán, J. (1987) 'La alquimia española del Siglo de Oro', in *Cinco tratados españoles de Alquimia* (Madrid: Tecnos), pp. 114–16.
Espósito, R. (2006) *Bíos: biopolítica y filosofía* (Buenos Aires: Amorrortu).
Ettinghausen, H. (ed.) (1995) *Noticias del siglo XVII: relaciones españolas de sucesos naturales y sobrenaturales* (Barcelona: Puvill Libros).
Falta, W. (1930) *Tratado de las enfermedades de las glándulas de secreción interna* (Barcelona: Labor).
Fausto-Sterling, A. (1997) 'How to build a man', in V. A. Rosario (ed.), *Science and Homosexualities* (London and New York: Routledge), pp. 219–25.

Fausto-Sterling, A. (2000) *Sexing the Body: Gender Politics and the Construction of Sexuality* (New York: Basic Books).
Ferguson, A. (1991) 'Androgyny as an ideal for human development', in A. Ferguson (ed.), *Sexual Democracy: Women, Oppression, and Revolution* (Boulder, CO, and Oxford: Westview Press), pp. 189–216.
Fernández, P. (1996) 'Moral social y sexual en el siglo XIX: la reivindicación de la sexualidad femenina en la novela naturalista radical', in I. M. Zavala (ed.), *Breve historia feminista de la literatura española (en lengua castellana)*, vol. 3, *La mujer en la literatura española: Modos de representación desde el siglo XVIII a la actualidad* (Barcelona and San Juan: Anthropos and Universidad de Puerto Rico), pp. 81–113.
Fernández del Valle, J. (1796–7) *Cirugía forense, general y particular*, 3 vols (Madrid: Imprenta de Aznar).
Fletcher, R. (1998) *Moorish Spain* (London: Phoenix).
Foderé, F. E. (1801–3 [1798]) *Las leyes ilustradas por las ciencias físicas ó tratado de medicina legal y de higiene pública*, 8 vols (Madrid: Imp. de la Administración del Real Arbitrio de Beneficencia).
Folguera Crespo, P. (1997) 'El franquismo: el retorno a la esfera privada (1939–1975)', in E. Garrido (ed.), *Historia de las mujeres en España* (Madrid: Síntesis), pp. 527–48.
Fontes Gil, R., L. Lorente Fernández, M. Jiménez Casado and P. Arnal Arambillet (1960) 'Un caso de seudohermafroditismo femenino por hiperplasia suprarrenal congénita (síndrome andrenogenital)', *Revista Clínica Española*, 76(5), 312–16.
Foucault, M. (1980) *Herculine Barbin, Being the Recently Discovered Memoirs of a Nineteenth-Century French Hermaphrodite*, trans. R. McDougall (New York: Pantheon Books).
Foucault, M. (1989) *The Birth of the Clinic: An Archaeology of Medical Perception*, trans. A. M. Sheridan (London: Routledge).
Foucault, M. (1990a) *The History of Sexuality*, vol. 1, *An Introduction*, trans. R. Hurley (Harmondsworth: Penguin).
Foucault, M. (1990b) 'La vida de los hombres infames', in *La vida de los hombres infames: Ensayos sobre desviación y dominación* (Madrid: La Piqueta), pp. 175–202.
Foucault, M. (1994a) 'Chronologie', in *Dits et écrits 1954–1988*, vol. 1 (Paris: Gallimard), 54.
Foucault, M. (1994b) 'Le vrai sexe', in *Dits et écrits 1954–1988*, vol. 4 (Paris: Gallimard), pp. 115–23.
Foucault, M. (1997) *'Il faut défendre la société' : cours au Collège de France, 1976* (Paris: Seuil).
Foucault, M. (2001) *Los anormales: curso del Collège de France (1974–1975)* (Madrid: Akal Universitaria).
Fragoso, J. (1570) *Erotemas chirúrgicos* (Madrid: Sebastián Yánez).
Fragoso, J. (1627 [1581]) *Cirugía universal* (Madrid: Viuda de Alonso Martín).
Frattale, L. (1989) 'Introducción', in José Antich, *Andrógino: poema [1904]* (Madrid: Tecnos), pp. 9–42.

Fuchs, B. (1996) 'Border crossings: transvestism and passing in "Don Quijote"', *Cervantes: Bulletin of the Cervantes Society of America*, 16(2), 4–28.

Galcerán, A. (1877) 'Hipospadias: hermafrodismo (un caso)', *RMCP*, primera parte. sección científica, 177–8.

Garber, M. (1991) 'The chic of araby: transvestism, transsexualism and the erotics of cultural appropriation', in J. Epstein and K. Straub (eds), *Body Guards: The Cultural Politics of Gender Ambiguity* (New York and London: Routledge), pp. 223–47.

García-Ballester, L. (1984) *Los moriscos y la medicina: un capítulo de la medicina y la ciencia marginadas en la España del siglo XVI* (Barcelona: Labor).

García-Ballester, L. (2002) 'The circulation and use of medical manuscripts in Arabic in sixteenth-century Spain', in J. Arrizabalaga, M. Cabré, L. Cifuentes and F. Salmón (eds), *Galen and Galenism: Theory and Medical Practice from Antiquity to the European Renaissance* (Aldershot and Burlington, UT: Ashgate), pp. 183–99.

García Carrero, P. (1605) *Disputationes Medicae Super Libros Galeni de Locis Affectis et de Aliis Morbis ab eo Relictis* (Alcalá de Henares: Ex Officina J. Sánchez Crespo).

García Font, J. (1976) *Historia de la alquimia en España* (Madrid: Editora Nacional).

García-Nieto París, M. C. (2000) 'Trabajo y oposición popular de las mujeres durante la dictadura franquista', in G. Duby and M. Perrot (eds), *Historia de las mujeres*, vol. 5, *El siglo XX* (Madrid: Taurus), pp. 722–35.

García Orcoyen, J. (1929) 'Concepto actual de la hormona sexual femenina', *AMCE*, 30(16), 500–5.

García Orcoyen, J. (1932) 'Hormonas femeninas', *AMCE*, 35(19), 380.

García Triviño, F. (1929) 'A propósito de la titulación biológica de las hormonas sexuales femeninas', *AMCE*, 30(20), 609–11.

Garza, F. (2002) *Quemando mariposas: sodomía e imperio en Andalucía y México, siglos XVI–XVII* (Barcelona: Laertes).

Gilman, S. L. (1994) 'Sigmund Freud and the sexologists: a second reading', in R. Porter and M. Teich (eds), *Sexual Knowledge, Sexual Science: The History of Attitudes to Sexuality* (Cambridge: Cambridge University Press), pp. 323–48.

Gil y Gil, C. (1929) 'Titulación biológica de la hormona sexual femenina', *AMCE*, 30(13), 409–11.

Gimeno Cabañas, A. (1931) 'Algunas consideraciones sobre un pretendido caso de hermafrodismo', *SM*, 4032, 313.

Glick, T. F. (1976) 'On the diffusion of a new specialty: Marañón and the "crisis" of endocrinology in Spain', *Journal of the History of Biology*, 9(2), 287–300.

Glick, T. F. (2005) 'Marañón, intersexuality and the construction of gender in 1920s Spain', *Cronos: Cuadernos Valencianos de Historia de la Medicina y de la Ciencia*, 8(1), 121–37.

Goldschmidt, R. (1917) 'Intersexuality and the endocrine aspect of sex', *Endocrinology* (Philadelphia), 1, 433–56.

González Duro, E. (1996) *Historia de la locura en España*, vol. 3, *Del reformismo del siglo XIX al franquismo* (Madrid: Temas de Hoy).

González Galván, J. (1940) 'Biología poética del amor', *Actualidad Médica*, 26(189), 73.

Granjel, L. S. (1967) 'La obra de Álvarez de Miraval', in *Médicos españoles* (Salamanca: Universidad de Salamanca), pp. 93–116.

Granjel, L. S. (1980) *La medicina Española renacentista* (Salamanca: Universidad de Salamanca).

Granjel, L. S. (2001) 'El médico Andrés Laguna', in J. L. García Hourcade and J. M. Moreno Yuste (eds), *Andrés Laguna: humanismo, ciencia y política en la Europa renacentista* (Valladolid: Junta de Castilla y León, Consejería de Educación y Cultura), pp. 11–16.

Grote (1938) 'Tratamiento general de la insuficiencia germinal de la mujer', *Actualidad Médica*, 24(161), 198–9.

Gutiérrez de Torres, A. (1952 [1524]) *El sumario de las maravillosas y espantables cosas que en el mundo han acontecido* (Madrid: Real Academia de la Lengua Española).

Hacking, I. (1999) *The Social Construction of What?* (Cambridge, MA and London: Harvard University Press).

Hall, D. L. (1976) 'The critic and the advocate: contrasting British views on the state of endocrinology in the early 1920s', *Journal of the History of Biology*, 9(2), 269–85.

Hall, D. L. and T. F. Glick (1976) 'Endocrinology: a brief introduction', *Journal of the History of Biology*, 9(2), 229–33.

Hampson, J. G. (1955) 'Hermaphroditic genital appearance, rearing and eroticism in hyperadrenocorticism', *Bulletin of the Johns Hopkins Hospital*, 96, 265–73.

Harding, S. (1986) *The Science Question in Feminism* (Ithaca, NY and London: Cornell University Press).

Haro García, F. (1926) Review of Marañón, *Tres ensayos sobre la vida sexual*, in *Actualidad Médica* (Granada), 4(21), 185–9.

Haro García, F. (1929) '"Los estados intersexuales en la especie humana", resumen de la obra del Dr. Marañón', in *Actualidad Médica* (Granada), 9(53), 313–19.

Hausman, B. (1995) *Changing Sex: Transsexualism, Technology, and the Idea of Gender* (Durham, NC, and London: Duke University Press).

Hausman, B. L. (1992) 'Demanding subjectivity: transsexualism, medicine and the technologies of gender, *Journal of the History of Sexuality*, 3(2), 270–302.

Herdt, G. (1996) *Third Sex, Third Gender: Beyond Sexual Dimorphism in Culture and History* (New York: Zone Books).

Herrn, R. (1995) 'On the history of biological theories of homosexuality', in J. P. De Cecco and D. A. Parker (eds), *Sex, Cells, and Same-Sex Desire* (New York: Howarth Press), pp. 31–56.

Hofman, E. (1882) *Elementos de medicina legal y toxicología* (Madrid: Imprenta de Enrique Teodoro).

Hofman, E. (1891) *Tratado de medicina legal*, 2 vols, vol. 2 (Madrid: Administración de la Revista de Medicina y Cirugía Prácticas).

Homero Arjona, J. (1937) 'El disfraz varonil en Lope de Vega', *Bulletin Hispanique*, 39, 120–45.

Huerta, L. (1929) 'El marañonismo y la intersexualidad', *Estudios*, 69, 9–12.
Huertas, R. and C. Ortiz (eds) (1997) *Ciencia y fascismo* (Madrid: Doce Calles).
Huertas García-Alejo, R. (1990) 'El concepto de "perversión sexual" en la medicina positivista', *Asclepio*, 42(2), 89–100.
Huarte de San Juan, J. (1977 [1575]) *Examen de ingenios para las ciencias* (Madrid: Editora Nacional).
Iglesias Aparicio, P. (2007) *Construcción del sexo y género desde la Antigüedad al siglo XIX* (Vigo: Universidad de Vigo): http://webs.uvigo.es/pmayobre/textos/pilar_iglesias_aparicio/tesis_doctoral/cap2_construccion_de_sexo_y_genero_desd e_%20la_edad_media.doc.
Iglesias y Díaz, M. (1892) 'Sociedades científicas: Real Academia de la Medicina, sesión del 11 de mayo de 1892', *SM*, 2019, 574.
Inamoto, K. (1992) 'La mujer vestida de hombre en el teatro de Cervantes', *Cervantes: Bulletin of the Cervantes Society of America*, 12(2), 137–43.
Irueste, J. (1908) 'Impotencia funcional por vicio de conformación genital congénito', *RMCP*, 79, 394–6.
Jacquart, D. and C. Thomasset (1989) *Sexualidad y saber médico en la Edad Media* (Barcelona: Labor).
Jacquart, D. (1997) 'Influence de la médicine arabe en Occident médiéval', in Roshdi Rashed (ed.), *Histoire des sciences arabes*, vol. 3, *Technologie, alchimie et sciences de la vie* (Paris: Éditions du Seuil), pp. 213–32.
Jagoe, C. (1994) *Ambiguous Angels: Gender in the Novels of Galdós* (Berkeley– Los Angeles and London: University of California Press).
Jiménez de Asúa, L. (1934) 'Aspecto jurídico de la maternidad consciente', in E. Noguera and L. Huerta (eds), *Genética, eugenesia y pedagogía sexual*, 2 vols, vol. 1 (Madrid: Javier Morata), pp. 333–41.
J. M. M. (1888) [No title], *RMCP*, 22, 468–9.
Jones, H. W. Jr. and W. W. Scott (1958) *Hermaphroditism: Genital Abnormalities and Related Endocrine Disorders* (Baltimore, MD: Williams & Wilkins).
Juaristi, V. (1943) 'El síndrome de Fröelich, transitorio', *Gaceta Médica Española*, 8–10.
Juliá, S. (2004) *Historia de las dos Españas* (Madrid: Taurus).
Kappler, C. (1986) *Monstruos, demonios y maravillas a fines de la Edad Media* (Madrid: Akal).
Karkazis, L. (2008) *Fixing Sex: Intersex, Medical Authority, and Lived Experience* (Durham, NC and London: Duke University Press).
Kaufman, C. (1938) 'Tratamiento de la insuficiencia ovárica con hormonas genitales', *Actualidad Médica*, 24(101), 196.
Kessler, S. J. (1998) *Lessons from the Intersexed* (New Brunswick, NJ, and London: Rutgers University Press).
Kirkpatrick, S. (1989) *Las Románticas: Women Writers and Subjectivity in Spain, 1835–1850* (Berkeley–Los Angeles and London: University of California Press).
Koyré, A. (1971) *Mystiques, spirituels, alchimistes du XVIe siècle allemand* (Paris: Gallimard).
Kuhn, T. (1970) *The Structure of Scientific Revolutions* (Chicago: Chicago University Press).

Labanyi, J. (2000) *Gender and Modernization in the Spanish Realist Novel* (Oxford: Oxford University Press).
Lalinde Abadía, J. (1986) 'La indumentaria como símbolo de discriminación jurídico social', *Anuario de Historia del Derecho Español*, 53, 583–601.
Laqueur, T. (1990) *Making Sex: Body and Gender from the Greeks to Freud* (Cambridge, MA and London: Harvard University Press).
Laqueur, T. (2003) 'Sex in the Flesh', *Isis*, 94, 300–6.
Latour, B. (1987) *Science in Action: How to Follow Scientists and Engineers through Society* (Cambridge, MA: Harvard University Press).
Latour, B. and S. Woolgar (1986) *Laboratory Life: The Construction of Scientific Facts* (Princeton, NJ: Princeton University Press).
Le Goff, J. (1992) *The Medieval Imagination*, trans. Arthur Goldhammer (Chicago and London: University of Chicago Press).
Legrand du Saulle, H., G. Berryer and G. Pouchet (1898) *Tratado de medicina legal de jurisprudencia médica y de toxicología*, vol. 2 (Madrid: Librería de Hernando y Compañía).
Libis, J. (2001) *El mito del andrógino* (Madrid: Siruela).
Litvak, L. (1979) *Erotismo fin de siglo* (Barcelona: Antoni Bosch).
Lobera de Avila, L. (1551) *Libro de regimiento de la salud y de la esterilidad de los hombres y mujeres* (Valladolid: Sebastián Martínez).
López Escoriaza and García Orcoyen (1932) 'Función ovárica y hermafroditismo', *AMCE*, 35(26), 519–20.
López Peláez (1916) 'Del estado civil de los hermafroditas', *SM*, 3278, 650.
López Piñero, J. M., T. F. Glick, V. Navarro Brótons and E. Portela Marco (1983a) 'García Carrero, Pedro', in *Diccionario Histórico de la Ciencia Moderna en España*, vol. 1 (Barcelona: Península), p. 374.
López Piñero, J. M., T. F. Glick, V. Navarro Brótons and E. Portela Marco (1983b) 'Bravo de Sobremonte', in *Diccionario Histórico de la Ciencia Moderna en España*, vol. 1 (Barcelona: Península), p. 134.
López Piñero, J. M., T. F. Glick, V. Navarro Brótons and E. Portela Marco (1983c) 'Mercado, Luis', in *Diccionario Histórico de la Ciencia Moderna en España*, vol. 2 (Barcelona: Península), pp. 56–9.
López Piñero, J. M., T. F. Glick, V. Navarro Brótons and E. Portela Marco (1983d) 'Valles, Francisco', in *Diccionario Histórico de la Ciencia Moderna en España*, vol. 2 (Barcelona: Península), p. 392.
Lozano y Ponce de León (1908) 'Ectopia testicular doble: Curación', *RMCP*, 80, 189–91.
McKendrick, M. (1974) *Women and Society in the Spanish Drama of the Golden Age: A Study of the 'Mujer Varonil'* (Cambridge: Cambridge University Press).
McVaugh, M. R. (1993) *Medicine before the Plague: Practitioners and their Patients in the Crown of Aragon, 1283–1345* (Cambridge: Cambridge University Press).
Maestre (1924) 'Un caso de hermafrodismo', *SM*, 3689, 183–4.
Magnan, V., and S. Pozzi (1911) 'Inversion du sens génital chez un pseudo-hermaphrodite féminin: sarcome de l'ovaire gauche opéré avec succès', *Bulletin de l'Académie de Médicine*, 3rd ser., 65(8), 223–59.
Mak, G. (2005) '"So we must go behind even what the microscope can

reveal": the hermaphrodite's "self" in medical discourse at the start of the twentieth century', *GLQ: A Journal of Lesbian and Gay Studies*, 11(1), 65–94.

Mangini, S. (2001) *Las modernas de Madrid: las grandes intelectuales españolas de la vanguardia* (Barcelona: Península).

Marañón, G. (1926a) 'La educación sexual y la diferenciación sexual', *Generación Consciente*, 31, 15–18.

Marañón, G. (1926b) 'La educación sexual y la diferenciación sexual', *Generación Consciente*, 32, 42–5.

Marañón, G. (1928) 'Un caso de intersexualidad de tipo hermafrodítico', *MI*, 22(572), 381.

Marañón, G. (1929) prologue to H. [*sic*, for A.] Hernández-Catá, *El Ángel de Sodoma* (Valparaíso: 'El Callao'), pp. 9–19.

Marañón, G. (1931a) 'Intersexualidad histológica e intersexualidad química', *SM*, 4069, 587–8.

Marañón, G. (1931b) 'Consideraciones sobre un caso de eunucoidismo', *AMCE*, 24(44), 1013.

Marañón, G. (1931c) 'Virilismo Postgravídico', *SM*, 88, 507–8.

Marañón, G. (1932a) 'Investigación de las hormonas hipofisogenitales en un nuevo caso de homosexualidad', *MI*, 26(746), 280.

Marañón, G. (1932b) 'Acerca del problema de la intersexualidad', *SM*, 4082, 243–7.

Marañón, G. (1936) 'La endocrinología y la ciencia penal', prologue, in Q. Saldaña, *Nueva criminología* (Madrid: M. Aguilar), pp. 7–18.

Marañón, G. (1950 [1943]) *Manual de diagnóstico etiológico* (Madrid: Espasa Calpe).

Marañón, G. (1966a [1922]) 'Estado actual de la doctrina de las secreciones internas', in *Obras completas*, vol. 2 (Madrid: Espasa-Calpe), pp. 9–89.

Marañón, G. (1966b [1934]) 'Las hormonas sexuales', prologue to B. Zondek, *Las hormonas del ovario y del lóbulo anterior de la hipófisis*, in *Obras completas*, vol. 1 (Madrid: Espasa Calpe), pp. 133–6.

Marañón, G. (1966c) 'La endocrinología y la ciencia penal', in *Obras completas*, vol. 1 (Madrid: Espasa Calpe), pp. 569–75.

Marañón, G. (1967a [1920]) 'Biología y feminismo', in *Obras completas*, vol. 3 (Madrid: Espasa-Calpe), pp. 9–33.

Marañón, G. (1967b [1925]) 'Sexo, trabajo y deporte', in *Obras completas*, vol. 3 (Madrid: Espasa-Calpe), pp. 95–112.

Marañón, G (1967c) 'Los estados intersexuales en la especie humana', in *Obras completas*, vol. 3, *Conferencias* (Madrid: Espasa-Calpe), pp. 155–85.

Marañón, G. (1968a [1928]) 'Nuevas ideas sobre el problema de la intersexualidad y sobre la cronología de los sexos', in *Obras completas*, vol. 4, *Artículos y otros trabajos* (Madrid: Espasa-Calpe), pp. 165–83.

Marañón, G. (1968b) 'Influencia de las secreciones internas en la evolución de la sexualidad', in *Obras completas*, vol. 4, *Artículos y otros trabajos* (Madrid: Espasa-Calpe), pp. 187–201.

Marañón, G. (1968c [1959]) 'Un caso de homosexualidad femenina con sexo cromático masculino', in *Obras completas*, vol. 4, *Artículos y otros trabajos*, (Madrid: Espasa-Calpe), pp. 1043–5.

Marañón, G. (1972 [1930]) *La evolución de la sexualidad y los estados intersexuales*, in *Obras completas*, vol. 8 (Madrid: Espasa-Calpe), pp. 499–710.
Marañón, Pardo and Gómez Acebo (1932) 'Un caso de hipervirilismo seguido de eunucoidismo', *AMCE*, 35(44), 897–8.
Marañón and Planelles (1932) 'Hormonas genitales y eunucoidismo', *AMCE*, 35(7), 128–9.
Marc, C. C. H. (1817) 'Hermaphrodite', in *Dictionnaire des sciences médicales par une société de médecins et de chirurgiens*, vol. 21, *HEM–HUM* (Paris: C. L. F. Panckoucke), pp. 86–121.
Mariani (1880) 'Hermafrodismo asimétrico', *RMCP*, 6, 337.
Mariani (1886) 'Un caso de hermafrodismo', *RMCP*, 18, 41–2.
Martín del Río (1991 [1599–1600]) *La magia demoníaca: libro II de las disquisiciones mágicas* (Madrid: Hiperión).
Martínez Pérez, J. (1988) 'Sobre la incorporación del método experimental a la medicina legal española: el estudio de las manchas de sangre en la obra de Lecha-Marzo', in M. E. Piñero *et al.* (eds), *Estudios sobre historia de la ciencia y de la técnica (Actas del IV Congreso de la Sociedad Española de Historia de las Ciencias y de las Técnicas)* (Valladolid: Junta de Castilla y León), pp. 833–44.
Mas Collellmir, J. (1932) 'Un caso de pseudohermafroditismo', *Revista Médica de Barcelona*, 18, 12–15.
Mata, P. (1844) *Vade mecum de medicina y cirugía legal*, vol. 1 (Madrid: Imprenta Calle de Padilla).
Mata, P. (1857) *Tratado de medicina y cirugía legal teórica y práctica*, vol. 1 (Madrid: Bailly-Baillière).
Mata, P. (1874) *Tratado de medicina y cirugía legal teórica y práctica*, 4 vols (Madrid: Bailly-Baillière).
Mata, P. (1903) *Tratado teórico-práctico de medicina legal y toxicología* (Madrid: Bailly-Baillière e Hijos).
Mauss, M. (1992 [1934]), 'Techniques of the body', in J. Crary and S. Kwinter (eds), *Incorporations* (New York: Zone), pp. 455–77.
Medvei, V. C. (1982) *A History of Endocrinology* (Lancaster: MTP Press).
Meeks, W. A. (1974) 'The image of the androgyne: some uses of a symbol in earliest Christianity', *History of Religions*, 13(3), 165–208.
Mejía, N. (2006) *Transgenerismos: Una experiencia transexual desde la perspectiva antropológica* (Barcelona: Ed. Bellaterra).
Mercado, L. (1579) *De Mulierum Affectionibus* (Valladolid: Diego Fernández de Córdoba).
Mexía, P. (1989 [1540]) *Silva de Varia Lección* (Madrid: Cátedra).
Meyerowitz, J. (2002) *How Sex Changed: A History of Transsexuality in the United States* (Cambridge, MA and London: Harvard University Press).
Mira, A. (2004) *De Sodoma a Chueca: Una historia cultural de la homosexualidad en España en el siglo XX* (Madrid: Egales).
Money, J. (1955) 'Hermaphroditism, gender and precocity in hyperadrenocorticism: psychologic findings', *Bulletin of the Johns Hopkins Hospital*, 96, 253–64.
Money, J., J. G. Hampson and J. L. Hampson (1955) 'Hermaphroditism: recommendations concerning assignment of sex, change of sex, and

psychologic management', *Bulletin of the Johns Hopkins Hospital*, 97, 284–300.

Montaña de Monserrate, B. (1551) *Libro de la anathomía del hombre* (Valladolid: S. Martínez).

Morales, A. (1923) 'Hipospadias', *SM*, 3626, 549–50.

Morales Pérez, A. (1906) 'Hipospadias de tercer grado aparentando un hermafrodismo', *Revista Médica de Sevilla*, 67(9), 257–61.

Morel d'Arleux, A. (1996) 'Las "Relaciones de Hermafroditas": dos ejemplos diferentes de una misma manipulación ideológica', in M. Cruz García de Enterría, *et al.*, *Las relaciones de Sucesos en España (1500–1750): Actas del Primer Coloquio Internacional (Alcalá de Henares, 8, 9 y 10 de junio de 1995)* (Alcalá de Henares and Paris: Pub. Universidad Alcalá de Henares and Pub. Sorbonne), pp. 261–71.

Morer Fargas, F. (1962) 'Los síndromes de intersexualidad cromosómica', *Medicina Clínica*, 38(3), 165–70.

Morland, I. (2001) 'Is intersexuality real?', *Textual Practice*, 15(3), 527–47.

Moscucci, O. (1991) 'Hermaphroditism and sex difference: the construction of gender in Victorian England', in M. Benjamin (ed.), *Science and Sensibility: Gender and Scientific Enquiry, 1780–1945* (Oxford and Cambridge, MA: Blackwell), pp. 174–99.

Mosse, G. L. (1994) 'Masculinity and the decadence', in R. Porter and M. Teich (eds), *Sexual Knowledge, Sexual Science: The History of Attitudes to Sexuality* (Cambridge: Cambridge University Press), pp. 251–66.

Mullins, N. (1972) 'The development of a scientific specialty: the Phage Group and the origins of molecular biology', *Minerva*, 10, 51–82.

Muñoyerro Pretel, A. (1931) 'Un caso de pseudohermafroditismo masculino o androginoidismo', *AMCE*, 34(22), 505–6.

Muñoyerro Pretel, A. (1936a) 'Un caso de pseudohermafroditismo femenino o ginandroidismo', *AMCE*, 39(3), 94–5.

Muñoyerro Pretel, A. (1936b) 'Seudohermafroditismo femenino o ginandroidismo', *Gaceta Médica Española*, 114, 313–14.

Muñoyerro Pretel, A. (1936c) 'Seudohermafroditismo femenino o ginandroidismo', *AMCE*, 39(1), 21–2.

Nash, M. (1995) *Defying Male Civilization: Women in the Spanish Civil War* (Denver, CO: Arden Press).

Nash, M. (2000) 'Maternidad, maternología y reforma eugénica en España 1900–1939', in G. Duby and M. Perrot (eds), *Historia de las mujeres*, vol. 5, *El siglo XX* (Madrid: Taurus), pp. 689–92.

Navarro Cánovas, B. (1904) 'Una forma rara de epispadias: epispadias del balano y su formación', SM, 2636, 419.

Nederman, C. J., and J. True (1996) 'The third sex: the idea of the hermaphrodite in twelfth-century Europe', *Journal of the History of Sexuality*, 6(4), 497–517.

Nevado Requena, R. (1906) 'El hipospadias y hermafroditismo aparente', *SM*, 2740, 373–75.

Nielfa Cristóbal, G. (2003) 'El debate feminista durante el franquismo', in G. Nielfa Cristóbal (ed.), *Mujeres y hombres en la España franquista: Sociedad, economía, política, cultura* (Madrid: Editorial Complutense), pp. 275–8.

Nieremberg, J. E. (1643 [1638]) *Curiosa y oculta filosofía* (Madrid: Imprenta Real).
Nieto, J. A. (ed.) (1998) *Transexualidad, transgenerismo y cultura: Antropología, identidad y género* (Madrid: Talasa).
Nieto, J. A. (2004) 'Globalización y transgenerismo en el área del Pacífico', *Revista Española del Pacífico*, 16, 191–221.
Novoa Santos, R. (1916) *Manual de patología general*, 3 vols, vol. 1 (Santiago de Compostela: Tipografía El Eco de Santiago).
Nye, R. (1989) 'Sex difference and male homosexuality in French medical discourse, 1830–1930', *Bulletin of the History of Medicine*, 63(1), 32–51.
Nye, R. A. (1991) 'The history of sexuality in context: national sexological traditions', *Science in Context*, 4(2), 387–406.
Oakley, A. (1972) *Sex, Gender and Society* (London: Temple Smith).
Oosterhuis, H. (2000) *Stepchildren of Nature: Krafft-Ebing, Psychiatry, and the Making of Sexual Identity* (Chicago and London: University of Chicago Press).
Orfila, M. (1847) *Tratado de medicina legal*, 4 vols (Madrid: Imprenta de Don José María Alonso).
Ortega Núñez, A., M. Fernández Fernández and A. Casabón (1960) 'Seudohermafroditismo masculino hereditario', *Revista Clínica Española*, 77(6), 386–92.
Ortega y Gasset, J. (1983 [1906]) 'La ciencia romántica', in *Obras completas*, vol. 1, *Artículos (1902–1913)* (Madrid: Alianza and Revista de Occidente), pp. 38–43.
Ortiz, T. (1993) 'El discurso médico sobre las mujeres en la España del primer tercio del siglo veinte', in M. T. López Beltrán (ed.), *Las mujeres en Andalucía: Actas del 2° encuentro interdisciplinar de estudios de la mujer en Andalucía*, vol. 1 (Malaga: Diputación Provincial de Málaga), pp. 107–38.
Otero Carvajal, L. E. (ed.) (2006) *La destrucción de la ciencia en España: Depuración universitaria en el franquismo* (Madrid: Editorial Complutense).
Oudshoorn, N. (1990) 'Endocrinologists and the conceptualization of sex, 1920–1940', *Journal of the History of Biology*, 23(2), 163–86.
Oudshoorn, N. (1994) *Beyond the Natural Body: An Archeology of Sex Hormones* (London and NewYork: Routledge).
Ovid (2004) *Metamorphoses*, trans. David Raeburn (London: Penguin).
Pallardo Sánchez, L. F., M. Santiago Corchado, A. Cerdán Vallejo and M. Nistal (1972) 'Algunos aspectos del eunocoidismo hipogonadotrópico', *Medicina Clínica*, 59, 390–6.
Pardo, López Escoriaza and Conde Gargollo (1932) 'Nota previa sobre asimetría de los caracteres sexuales', *AMCE*, 35(10), 189–90.
Paré, A. (1982) *On Monsters and Marvels*, trans. J. L. Pallister (Chicago and London: University of Chicago Press).
Park, K. (1985) *Doctors and Medicine in Early Renaissance Florence* (Princeton, NJ: Princeton University Press).
Park, K. (1997) 'The rediscovery of the clitoris: French medicine and the tribade, 1570–1620', in D. Hillman and C. Mazzio (eds), *The Body in Parts: Fantasies on Corporeality in Early Modern Europe* (New York and London: Routledge), pp. 171–93.

Park, K. (2000) 'Una historia de la admiración y del prodigio', in A. Lafuente and J. Moscoso (eds), *Monstruos y seres imaginarios en la Biblioteca Nacional* (Madrid: Ministerio de Educación y Cultura, Biblioteca Nacional), pp. 77–90.

Park, K. and L. Daston (1981) 'Unnatural conceptions: the study of monsters in sixteenth and seventeenth century France and England', *Past and Present*, 92, 20–54.

Park, K. and R. A. Nye (1991) 'Destiny is anatomy', *New Republic*, 53–7.

Pascual, S. (1920) 'Un caso de hermafrodismo', *MI*, 11(135), 181.

Passeron, J. C. (2006) *Le raisonnement sociologique: un espace non-poppérien de l'argumentation* (Paris: Albin Michel).

Paterson, A. K. G. (1993) 'Tirso de Molina and the androgyne: "El Aquiles" y "La Dama del Olivar"', *Bulletin of Hispanic Studies*, 70(1), 105–13.

Peña Regidor, P. (1941) 'Tiroides y atrofia ovárica', *Revista Española de Farmacología y Terapéutica*, 21, 86–7.

Peratoner, A. (1880) *Higiene y fisiología del amor en los dos sexos* (Barcelona: La Moderna Maravilla, Gran Casa Editorial).

Pérez de Moya, J. (1585) *Philosophía secreta* (Madrid: Francisco Sánchez).

Perez, J. P. and C. Cherizola (1864) 'Hermafrodismo', *La Crónica Médica*, 34, 74–9.

Pérez Ibáñez, M. J. (1997) *El humanismo médico del siglo XVI en la Universidad de Salamanca* (Valladolid: Universidad de Valladolid).

Perry, M. E. (1999) 'From convent to battlefield: cross-dressing and gendering the self in the New World of imperial Spain', in J. Blackmore and G. S. Hutcheson (eds), *Queer Iberia: Sexualities, Cultures and Crossings from the Middle Ages to the Renaissance* (Durham, NC: Duke University Press), pp. 394–419.

Peset Cervera (1879) 'Notable caso de pseudo-hermafrodismo', *RMCP*, 5, 418.

Peset, J. L. and M. Peset (1975) 'Estudio preliminar', *Lombroso y la escuela positivista italiana* (Madrid: CSIC), pp. 80–1.

Pinar, S. (2002) 'The emergence of modern genetics in Spain and the effects of the Spanish Civil War (1936–1939) on its development', *Journal of the History of Biology*, 35(1), 111–48.

Piulachs, P. (1943) 'Un caso de seudohermafroditismo', *Revista Clínica Española*, 10(3), 210–12.

Planelles (1932a) 'Bosquejo general: hormonas prehipofisiarias de acción genital', *AMCE*, 35(19), 380.

Planelles (1932b) 'Hormonas genitales en el hombre', *AMCE*, 35(19), 381.

Platero, Raquel (2008) 'Outstanding challenges in a post-equality era: the same-sex marriage and gender identity laws in Spain', *International Journal of Iberian Studies*, 21(1), 41–9.

Pliny (1942) *Natural History*, trans. H. Rackham, vol. 2, book 7 (London and Cambridge, MA: William Heinemann and Harvard University Press).

Polo Blanco, A. (2003) *Otras Madres de Mayo: diseño y planificación de la estructura del ámbito familiar en el primer franquismo* (Cadiz: Servicio de Publicaciones de la Universidad de Cádiz).

Polo Blanco, A. (2005) 'Biopolítica y gestión de poblaciones en el primer franquismo (1939–1945)', unpublished doctoral thesis (Cadiz: Universidad de Cádiz).

Polo Blanco, A. (2006) 'Gobierno de las poblaciones en el primer franquismo (1939–1945)' (Cadiz: Servicio de Publicaciones de la Universidad de Cádiz).

Pomata, G. (1982) 'Uomini menstruanti somiglianza e diferenza fra i sessi in Europa in età moderna', *Quaderni Storici*, 79(1), 51–103.

Pons Tortella, E. and M. Gállego Berenguer (1934) 'Nota sobre un caso de pseudohermafroditismo con error de sexo', *Revista Médica de Barcelona*, 129, 213–17.

Porter, R. (1985) 'The patient's view: doing medical history from below', *Theory and Society*, 14(2), 175–98.

Pozzi, S. (1893) *A Treatise on Gynaecology, Clinical and Operative*, vol. 3 (London: New Sydenham Society).

Pozzi, S. (n.d.), *Tratado de ginecología clínica y operatoria*, vol. 2 (Barcelona: José Espasa Editor).

Pulido, A. (1880) 'Lactancia paterna', *RMCP*, 7, 13–22.

Quental (1938) 'Contribución al estudio de la hormona sexual femenina en el macho normal o castrado', *Actualidad Médica*, 24(167), 456.

Raso, V. (1978) *La Pastora: el maqui hermafrodita* (Bilbao: Ediciones Albia).

Recasens, S. (1918) *Tratado de ginecología* (Madrid: Imprenta y Librería de Nicolás Moya).

Rehg, W. and J. Bohman (eds) (2001) *Pluralism and the Pragmatic Turn* (Cambridge, MA: MIT Press).

Resa, R. (1932) 'Un caso de ginecomastia doble', *AMCE*, 35(7), 129.

Revel, J. (1982) 'El historiador y los papeles sexuales', in various authors, *Familia y sexualidad en Nueva España* (Mexico City: FCE), pp. 53–4.

Reventós, J., A. García and C. Piqué (1991) *Historia de la medicina catalana durante el franquismo y sus consecuencias* (Madrid: Ministerio de Sanidad y Consumo).

Richards, M. (1998) *A Time of Silence: Civil War and the Culture of Repression in Franco's Spain, 1936–1945* (Cambridge: Cambridge University Press).

Riera Palmero, J. (2001) 'Andrés Laguna y el Galenismo renacentista', in J. L. García Hourcade and J. M. Moreno Yuste (eds), *Andrés Laguna: humanismo, ciencia y política en la Europa renacentista* (Valladolid: Junta de Castilla y León, Consejería de Educación y Cultura), p. 166.

Rivilla Bonet, J. (1695) *Desvíos de la naturaleza o tratado del origen de los monstruos* (Lima: Imprenta Real).

Robert y Yarzábal, B. (1882) 'Un caso de hermafrodismo relacionado con la rino-bronquitis espasmódica', *Anales de Obstetricia, Ginecología y Pediatría*, 2, 151–3.

Roca i Girona, J. (1996) *De la pureza a la maternidad: la construcción del género femenino en la postguerra española* (Madrid: Ministerio de Educación y Cultura).

Roca i Girona, J. (2003) 'Esposa y madre a la vez', in G. Nielfa Cristóbal (ed.), *Mujeres y hombres en la España franquista: sociedad, economía, política, cultura* (Madrid: Editorial Complutense), pp. 45–63.

Roda, E., De la Peña, A. and De la Peña, E. (1944) 'Seudohermafroditismo masculino', *Revista Clínica Española*, 13(3), 165–71.

Rodríguez Morini (1900) 'Ensayo de transplantación de un ovario humano', *Gaceta Médica Catalana*, 23(2), 66.

Romera Navarro, M. (1934) 'Las disfrazadas de varón en la comedia', *Hispanic Review*, 2, 269–86.

Ronzón, E. (1998) 'El médico Juan Sánchez Valdés de la Plata y su libro sobre el hombre: historia de una investigación', *El Basilisco*, 24, 63–84.

Rosario, V. A. (1997) *The Erotic Imagination: French Histories of Perversity* (New York and Oxford: Oxford University Press).

Roscoe, W. (1991) *The Zuni Man-Woman* (Alburquerque, NM: University of New Mexico).

Rose, N. (1994) 'Medicine, history and the present', in C. Jones and R. Porter (eds), *Reassessing Foucault: Power, Medicine and the Body* (London and New York: Routledge), pp. 48–72.

Rose, N. (1999) *Governing the Soul: The Shaping of the Private Self* (London: Free Association Books).

Rubin, G. (1975) 'The traffic in women: notes on the "political economy" of sex', in R. R. Reiter (ed.), *Toward an Anthropology of Women* (New York and London: Monthly Review Press), pp. 157–210.

Russett, C. E. (1989) *Sexual Science: The Victorian Construction of Womanhood* (Cambridge, MA, and London: Harvard University Press).

Salamanca Ballesteros, A. (2007) *Monstruos, ostentos y hermafroditas* (Granada: Universidad de Granada).

Sánchez Ron, J. M. (1999) *Cincel, martillo y piedra: Historia de la ciencia en España (siglos XIX y XX)* (Madrid: Taurus).

Sánchez Valdés de la Plata, J. (1598) *Crónica y historia general del hombre* (Madrid: Miguel Martínez).

Sánchez de Viana, P. (1589) *Anotaciones sobre los quinze libros de las Transformaciones, de Ovidio* (Valladolid: Diego Fernández de Córdoba).

Sanz Hermida, J. (1993) 'Aspectos fisiológicos de la dueña dolorida: la metamorfosis de la mujer en hombre', in *Actas del Tercer Coloquio Internacional de la Asociación de Cervantistas* (Barcelona: Anthropos), pp. 463–72.

Sanz Hermida, J. (1997) 'La literatura de problemas en España (siglos XVI y XVII)', unpublished doctoral thesis (Salamanca: University of Salamanca).

Sarabia (1908) 'Hipospadias de tercer grado, aparentando un hermafroditismo', *RMCP*, 79, 446–8.

Schatzki, T. R., K. Knorr Cetina and E. Von Savigny (eds) (2001) *The Practice Turn in Contemporary Theory* (London: Routledge).

Schiebinger, L. (1989) *The Mind has No Sex? Women in the Origin of Modern Science* (Cambridge, MA and London: Harvard University Press).

Schiebinger, L. (1993) *Nature's Body: Sexual Politics and the Making of Modern Science* (London: HarperCollins).

Schiebinger, L. (2000) *Feminism and the Body* (Oxford: Oxford University Press).

Schiebinger, L. (2003) 'Skelettestreit', *Isis*, 94, 307–13.

Scott, J. W. (1999 [1986]) 'Gender: a useful category of historical analysis', in *Gender and the Politics of History* (New York: Columbia University Press), pp. 28–50.

Searle, J. (1996) *The Construction of Social Reality* (London: Penguin).

Sebileau, P. and R. Pichevin (1902) *Tratado de cirugía clínica y operatoria*, trans. José Núñez Granés (Madrid: Hernando y Compañía).

Sengoopta, C. (2006) *The Most Secret Quintessence of Life: Sex, Glands, and Hormones, 1850–1950* (Chicago and London: University of Chicago Press).

Señor, J. C. (1953) 'J. Botella Llusiá y F. Nogales: Sobre el síndrome de seudahermafroditismo [*sic*] masculino con feminización total. – Acta Gin.; t.3; pág. 319; 1952', *Medicina Clínica*, 21(1), 73.

Serrallach, D. N. (1908) 'Las glándulas sexuales del hombre y su nueva fisiología', *RMCP*, 80, 5–18.

Simarro Puig, J., A. Otero Sánchez and J. Lluch Caralps (1942) 'Error en la determinación del sexo: hipospadias perineal. Algunas consideraciones sobre los seudo-hermafroditas', *Revista Clínica Española*, 6(1), 39–42.

Simonin, C. (1962 [1955]) *Medicina Legal Judicial* (Barcelona: Jims).

Solomon, M. (1999) 'Diseasing the sexual other in Francesc Eiximenis *Lo Llibre de les dones*', in J. Blackmore and G. S. Hutcheson (eds), *Queer Iberia: Sexualities, Cultures and Crossings from the Middle Ages to the Renaissance* (Durham, NC: Duke University Press), pp. 277–90.

Spink, M. S. and G. L. Lewis (1973) 'Introduction', in Albucasis, *On Surgery and Instruments* (London: Wellcome Institute of the History of Medicine), pp. vii–xv.

Steinberg, S. (2001) *La Confusion des sexes: La travestissement de la Renaissance à la Révolution* (Paris: Fayard).

Steuer, D. (2005) 'A book that won't go away: Otto Weininger's *Sex and Character*', in O. Weininger, *Sex and Character: An Investigation of Fundamental Principles*, trans. L. Löb (Bloomington and Indianapolis: Indiana University Press), pp. xi–xlvi.

Stolberg, M. (2003) 'A woman down to her bones: the anatomy of sexual difference in the sixteenth and early seventeenth centuries', *Isis*, 94, 274–99.

Thoinot, L. (1928) *Tratado de medicina legal, traducido, anotado y adicionado con referencia a la legislación española y americana por W. Coroleu, Secretario Perpetuo de la Real Academia de Medicina de Barcelona*, vol. 2 (Barcelona: Salvat Editores).

Tomás y Valiente, F. (1990) 'El crimen y pecado contra natura', in F. Tomás y Valiente et al., *Sexo Barroco y otras transgresiones modernas* (Madrid: Alianza Universidad), pp. 54–5.

Torre Blanco, J. (1929) 'Concepto biológico de la mujer', *AMCE*, 31(22), 528–35.

Ufer, J. (1960) *Hormonoterapia en ginecología: Fundamentos y Práctica* (Madrid: Alhambra).

Usandizaga. M. and G. Sánchez Lucas (1931) 'Hermafroditismo masculino', *AMCE*, 34(17), 373–78.

Valcárcel, A. (1991) *Sexo y filosofía: Sobre 'mujer' y 'poder'* (Barcelona: Anthropos).

Valentí Vivó, I. (1889) *Tratado de Antropología Médica y Jurídica*, vol. 1 (Barcelona: J. Roviralta; vol. 2, 1894).
Valverde de Amusco, J. (1556) *Historia de la composición del cuerpo humano* (Rome: J. A. de Salamanca & A. Lafrery).
Vandebosch, D. (2006) *Y no con el lenguaje preciso de la ciencia: La ensayística de Gregorio Marañón en la entreguerra española* (Geneva: Librairie Droz).
Vara López, R. and J. G. Sánchez-Lucas (1925) 'Sobre tres casos de pseudo-hermafroditismo: estudio histológico', *AMCE*, 224, 529–35.
Varela, J. and F. Álvarez-Uría (1989) *Sujetos frágiles: Ensayos de sociología de la desviación* (Madrid: FCE).
Various authors (1929) 'Sobre unos cronicones', *AMCE*, 30(8), 5–6.
Vázquez García, F. (1995) 'La exclusión del hermafrodita y la invención ilustrada del único sexo verdadero', in A. Romero Ferrer (ed.), *Actas del VI Encuentro de la Ilustración al Romanticismo: juego, fiesta y transgresión (1750–1850)* (Cadiz: Servicio de Publicaciones de la Universidad de Cádiz), pp. 645–53.
Vázquez García, F. (1999) 'La imposible fusión: claves para una genealogía del cuerpo andrógino', in D. Romero de Solís, J. Bosco Díaz-Urmeneta Muñoz and J. López-Lloret (eds), *Variaciones sobre el cuerpo* (Seville: Servicio de Publicaciones de la Universidad de Sevilla), pp. 217–35.
Vázquez García, F. (2001) 'Androginia y pensamiento esencial', *Culturas: Suplemento Cultural del Diario de Sevilla* (5 July), 6.
Vázquez García, F. (2008) 'Del hermafrodita al transexual: elementos para una genealogía del cuerpo sexuado (España siglos XVI–XX)', in N. Corral (ed.), *Prosa corporal* (Madrid: Talasa), pp. 75–97.
Vázquez García, F. and A. Moreno Mengíbar (1995a) 'Un solo sexo: invención de la monosexualidad y expulsión del hermafroditismo', *Daimón: Revista de Filosofía*, 11, 95–112.
Vázquez García, F. and A. Moreno Mengíbar (1995b) 'Del sexo verdadero al sexo veraz: Reyes Carrasco, un caso de hermafrodismo en el siglo XIX', *El Viejo Topo*, 84, 73–80.
Vázquez García, F. and A. Moreno Mengíbar (1997a) *Sexo y razón: una genealogía de la moral sexual en España (siglos XVI–XX)* (Madrid: Akal).
Vázquez García, F. and A. Moreno Mengíbar (1997b) 'El hermafrodita Reyes Carrasco: identidad sexual en la España del siglo XIX', *Historia 16*, 258, 30–6.
Vázquez García, F. and A. Moreno Mengíbar (2000) 'Hermafroditas y cambios de sexo en la España moderna', in A. Lafuente and J. Moscoso (eds), *Monstruos y seres imaginarios en la Biblioteca Nacional* (Madrid: Ministerio de Educación y Cultura, Biblioteca Nacional), pp. 91–103.
Velasco, S. (2001) 'Marimachos, hombrunas, barbados: the masculine woman in Cervantes', *Cervantes: Bulletin of the Cervantes Society of America*, 20(1), 69–78.
Verdeguer, A. E. (1928) 'El problema de la sexualidad', *SM*, 82, 210.
Veyne, P. (1984) *Cómo se escribe la historia: Foucault revoluciona la historia* (Madrid: Alianza Editorial).
Villarreal Casas, J. (1946) 'El ciclo unifásico como causa de esterilidad y su tratamiento con la gonadotropina de suero de yegua', *Ser*, 46, 42–5.

Viñas y Mey, C. and R. Paz (1949) *Relaciones histórico-geográfico-estadísticas de los pueblos de España hechas por iniciativa de Felipe II: Provincia de Madrid* (Madrid: CSIC, Instituto Balmes de Sociología), pp. 630–1.

Vital Aza, I. (1932) 'Resultados obtenidos en la clínica con la investigación de las hormonas genitales en los trastornos menstruales y en su tratamiento', *AMCE*, 35(19), 381.

Vital Aza, I. (1941) 'Esterilidad femenina por disfunción hormonal', *Semana Médica*, 145, 565–75.

Wanrooij, B. P. F. (1990) *Storia del pudore: la questione sessuale in Italia 1860–1940* (Venice: Marsilio Editori).

Weeks, J. (1989) *Sex, Politics and Society: The Regulation of Sexuality since 1800* (London and New York: Longman).

Weil, K. (1992) *Androgyny and the Denial of Difference* (Charlottesville, VA: University Press of Virginia).

Weininger, O. (2005 [1903]) *Sex and Character: An Investigation of Fundamental Principles*, trans. L. Löb (Bloomington and Indianapolis: Indiana University Press).

Wilson, D. (1993) *Signs and Portents: Monstrous Births from the Middle Ages to the Enlightenment* (London and New York: Routledge).

Wotan (1908) 'Un caso de hipospadias', *SM*, 2836, 234.

Wright, S. (2004) 'Gregorio Marañón and the "cult of sex": effeminacy and intersexuality in "The Psychopathology of Don Juan" (1924)', *Bulletin of Spanish Studies*, 81(6), 717–38.

Yáñez, T. (1878) *Lecciones de medicina legal y toxicología* (Madrid: Librería de Saturnino Calleja).

Yates, F. A. (2006) [1972] *The Rosicrucian Enlightenment* (London: Routledge & Kegan Paul).

Young, I. M. (2005) *On Female Body Experience: 'Throwing Like a Girl' and Other Essays* (Oxford: Oxford University Press).

Index

Acts of the Christian Martyrs 61
Águeda, St 59
Alba y López, R. 93–5, 96
Albertus Magnus 38
Alcalá de Henares 45, 47
Alcira 103
Aldaraca, Bridget 84
Alemán, Mateo 55
Alexander VI, Pope 57
Alexina B. 92, 99
Alfonso VI, king of Leon and Castile 54
Alfonso X, king of Castile 56
Alicante 105
Almuñécar 134
Álvarez de Miravall, Blas 42
Amsterdam 182
Angulo, Dr 131–2
Antoninus Pius, emperor of Rome 54
Aphrodite 1, 4, 6
Archivos de Medicina, Cirugía y Especialidades 182
Aristotelianism 8, 18, 38, 40, 42, 45, 46, 47, 50, 51, 165, 237
Aristotle 3, 5, 6–7, 17, 18, 32, 42, 43, 44, 45–6, 53
Arnal Arambillet, P. 232
Asas Manterota, Benita 148
Augustine, St 33, 35, 50, 54
Aulus Gelius 38
Avicenna 7, 38, 42

Baeza 40
Bauhin, G. 50
Bakhtin, Mikhail 34
Baloardo, Dr 168
Barbin, Herculine 2, 226
Barcelona 89, 98, 105, 109, 153, 181, 186, 193, 205

Barco y Gasca, A. J. del 237
Barnsby, Dr 128
Barragán, Dr 132
Bauer, J. 182, 187
Berlin 128, 139, 158, 182
Bernaldo de Quirós, María 150–1
Berryer, Georges 99–101, 102
'Biología y feminismo' (Marañón) 149
Birke, Lynda 11–12, 224
Bodies that Matter (Butler) 12
Böhme, Jakob 62
Botella Llusià, José 180, 197–8, 199, 200–4, 205, 206–7, 228
Bouis, J. 102
Brachfeld, F. Oliver 168
Bravo de Sobremonte, Gaspar 17, 47–9, 52, 237
Brian, E. 140
Briand, J. 99, 102
Brouardel, Paul 101
Brown-Séquard, C. E. 152, 153
Butenandt, Adolf 190
Butler, Judith 12

Cadden, Joan 6, 32
Cadiz 97, 98, 234
Calliphanes 37
Camacho Alejandre, F. 134–7
Canjayar 101
Canon (Avicenna) 42
Carcassonne, M. 92, 93
Carcavilla, Mauricio 230
Cardenal, León 143, 165–6
Cardeñosa 58, 59
Careaga, Señorita 151
Carlos II, king of Spain 55
Carlos III, king of Spain 47
Carranza, Alonso 17, 46–7, 48–9, 57
Carrasco, Dr 92

Casamiento entre dos Damas 59–60
Casper, J. L. 102
Catalonia 207
Chase, Cheryl 97, 233, 234, 235
Chaudé, Ernest 99, 102
Cherizola, Carlos 97
Chevalier, Julien 9
Cicero 53
Clement of Alexandria 54
Codina Castellví, Dr 128
Condado de Benavente 37
Consuelo S. 103
Córdoba 47
Counter-Reformation, the 35, 56, 57, 58
Covarrubias, Sebastián de 17
Cuadra, Dr 91
Curso de clínica general ... (De Letamendi) 111–12

D'Arsonval, Arsène 152
Darwin, Charles 9
Daston, Lorraine 6, 7, 8, 17, 32
De Beauvoir, Simone 11
De Burgos, Carmen 150
De Centellas, Luis 62
De Céspedes, Elena 2, 32, 226, 227, 231
De Civitate Dei (Augustine) 35, 54
De Conceptu et Generatione Hominis (Rueff) 54
De Erauso, Catalina 32–3, 57, 226
De Fuentelapeña, Antonio 17, 38–9, 52, 57, 237
De Fuentes, Alonso 38
De Generatione Animalium (Aristotle) 6, 17, 32, 44, 45–6
'De hermaphroditis' (Bravo de Sobremonte) 48
De la Cerda, Juan 37, 38
De la Peña, A. 194–7
De la Peña, E. 194–7
De León, Andrés 42, 55
De León, Fray Luis 84
De Letamendi, José 82, 111–12
De Mulierum Affectionibus (Mercado) 43–4, 50
De Peñaranda, Brígida 47, 226

De Pineda, Juan 38
De Torquemada, Antonio 17, 37, 38
De Torreblanca y Villalpando, Francisco 47
Descent of Man, The (Darwin) 9
Desvíos de la naturaleza ... (Rivilla Bonet) 49–50
Deutscher Zeitschrift für Chirugie 137
Díaz de Vivar, Rodrigo 54
Disputatio de Vera Humanu Partus Naturalis ... (Carranza) 46
Disputationes Medicae Super Libros Galeni (García Carrero) 17, 45
Disquisiciones Mágicas (Martín del Río) 39
Douday, José 102–3, 105
Dreger, Alice D. 4, 15–16, 19, 79, 80, 95, 108–9, 139, 224, 234, 236
Du Laurens, André 38–9, 40, 44–5, 46, 48, 50, 52

Eiximenis, Francesc de 61
Elder, Ruth 150
Elementos de medicina y cirugía ... (Peiró and Rodrigo) 86
Ellis, Henry Havelock 9, 199, 203
Enlightenment, the 83, 237
Erigena, John Scotus 62
España Médica 21, 93
Esquerdo, Dr 105
Etymologies (Isidore) 35
Evolución de la sexualidad y los estados intersexuales, La (Marañón) 155, 157, 159, 166, 196

Falange, the 198
Fausto-Sterling, Anne 224, 234
Felicitas, St 61
Felipe II, king of Spain 43
Felipe III, king of Spain 45
Felipe IV, king of Spain 47
Fellner, Ofried 154
Fernández, Fernanda 33, 57
Fernández del Valle, J. 237
Ficino, Marsilio 62
Foderé, F. M. 86, 87
Fontes Gil, R. 232

Index

Foucault, Michel 4, 8, 12, 16, 18, 29–30, 31, 56, 147
Fragoso, Juan 17, 42, 55
Franco, Francisco 20, 180, 230
Francoism 188, 189–90, 193, 194, 198–200, 206
Frank, Robert T. 158, 166, 167, 185
Freud, Sigmund 29
Fulgoso, B. 38

Gaceta Médica Catalana 152
Galen 3, 5, 6–7, 17, 38, 39, 40, 42, 45, 51, 52
Galenism 42, 43, 44, 45, 47–8, 51, 227, 236–7
Gállego Berenguer, M. 186–7
Ganivet, Angel 230
García Carrero, Pedro 17, 45–6, 47, 48, 49, 52, 53, 237
García Orcoyen, J. 182
Garma, Rafael 182
Garnier, Marie/Germain 2
Geddes, Patrick 149
Gender Identity Law 233, 236
Gender Trouble (Butler) 12
Gerona 98
Gimeno Cabañas, A. 166
Gley, E. 153
Glick, Thomas F. 81, 148, 152, 153
Gnostics, the 58, 62
Goldschmidt, Richard 156, 181, 182, 196, 231
Gómez Acebo, Dr 182
Goyanes, Dr 132
Granada 40, 57, 133
Gutíerrez de Torres, Alvar 54
Gynécologie et Obstétrique 140

Haeckel, Ernst 9
Hampson, Joan 19
Harding, S. 108
Havana 155
Henri IV, king of France 45
Hermaphrodites et la Chirugie, Les (Ombrédanne) 191–2
Hermaphroditus 1, 5
Hermes 1
Hermes Trismegistus 58

Herodotus 47
Hervas y Pandero, L. 237
Higiene del matrimonio (Monlau i Roca) 147
Hinkle, Curtis 233
Hippocrates 3, 4, 5, 6, 7, 8, 16, 17, 18, 30, 38, 39, 40, 42–3, 45, 47, 49, 51, 52, 227, 236–7
Historia Anatomica Humani Corporis (Du Laurens) 38, 39, 44, 52
Hoepke, Hermann 164
Hofman, Eduard R. von 109
Huarte de San Juan, Juan 17, 42, 55
Huelva 97, 98

Inquisition, the 2, 47
Institucíon Libre de Enseñanza 188
Institute of Clinical and Medical Research (Madrid) 205
Institute of Medical Pathology (Madrid) 152
Institutio Foeminae Christianae (Vives) 84
Intersexual Society of North America (ISNA) 233
Isidore, St 35, 50, 54

Jagoe, Catherine 84–5
Jardin de flores curiosas (De Torquemada) 17
Jiménez Casado, M. 232
Jiménez Díaz, Carlos 195, 205
Journal d'Urologie 140
Jumas, Justine 92

Katenstein, M. 136
Kessler, Suzanne 20
Kiernan, James 9
Kirkpatrick, Susan 84
Klebs, Theodore 15, 101, 106, 107, 108–9, 110, 113, 137, 138–9, 140, 141–2, 144, 158, 179, 194, 195, 225, 226, 231, 236, 237, 238
Klinefelter's syndrome 123, 206, 208, 236
Krabbe, Dr 158
Krause, Karl 84
Kuhn, T. 108

Labanyi, Jo 84
Lacassagne, A. 140
Lagouffe, Dr 140
Langdon-Brown, Walter 152
Laqueur, Ernst 179, 182, 183, 190
Laqueur, Thomas 4, 5–6, 29–30, 31, 32, 45, 53, 225
Lefort, Maria 89, 92
Legrand du Saulle, H. 92, 99–101, 102
Lex Repentudarum (Ulpian) 56–7
Leyes ilustradas por las ciencias físicas, Las ... (Foderé) 86
Liberada, St 58
Lima 49, 50
Lipschütz, A. 158, 167, 182, 187
Livy 38, 53
Lluch Caralps, J. 191–3, 194
Llull, Ramón 62
López Aydillo, Dr 122
López Escoriaza, Dr 182
Lorente Fernández, L. 232
Los Lagares 227
Lucar 129
Lucretius 53
Lusitano, Amato 38
Lycosthenes, Conrad 54

Madrid 47, 55, 57, 92, 122, 131, 162, 181, 186, 190, 195, 207, 226
Maestre, Dr 144
Magdalena S. 122–3
Malthusianism 238
Manual completo de medicina legal y toxicología (Briand, Bouis and Casper) 102
Marañón, Gregorio 9, 80, 114, 122–4, 126, 146, 147–50, 152–4, 155–61, 162–3, 164, 165, 166–8, 179, 181–3, 184–5, 186, 187, 188, 190, 191, 196, 199, 225, 226, 230, 231, 238
Marc, Charles Chrétien Henri 14–15, 78, 81, 87–8, 89, 90, 91–2, 93, 95, 99, 101, 102, 103, 106, 107, 110, 113, 125, 126–7, 140, 143, 226, 227, 238
Margarita, Maria 89, 92

Mariani, Juan Manuel 101
Martín de Azpilcueta, Navarro 57
Martín del Río 37, 38, 39–40, 47, 52, 237
Mas Collemir, J. 186
'Mass excretion of oestrogenic ("female") hormone in the urine of the stallion' (Zondek) 155
Mata i Fontanet, Pedro 85, 88–94, 95–6, 99, 102, 103, 110, 141
Mechanismus und Psychologie der Geschlechtsbestimmung (Goldschmidt) 156
Meckel, Johann Friedrich 238
Medicina Ibera, La 21, 140, 141
Meixner, Dr 137
Mercado, Luis 43–4, 45, 50, 51, 237
Metamorphoses (Ovid) 1
Mexía, Pedro 38
Mira, Emilio 188
Moebius, Paul Julius 9, 147, 149, 230
Moi, Toril 12
Mollá, Dr 141
Money, John 19, 180, 206, 229, 238
Monlau i Roca, Pedro 147
Montaigne, Michel de 38, 47
Montaña de Monserrate, Bernardino 42
Montpellier 45
Moore, K. L. 207–8
Morales Pérez, Antonio 132–3, 136, 231, 237
Morland, Iain 234
Morris's syndrome 236
Muñoyerro Pretel, A. 186, 187
Muñoz, María 40–1, 226

Nagel, Dr 112
Negrín, Juan 188
Neo-Aristotelians, the 48
Neoplatonists, the 62
Neugebauer, Franz 112, 137, 158–9, 163, 184, 186, 187
Nevado Requena, Rafael 129–31
'New ideas on the problem of intersexuality' (Marañón) 149
Nieremberg, Juan Eusebio 38, 47

Nogales 195
Novoa Santos, Roberto 113–14, 147, 230
Núñez Granés, Carlos 99

Oakley, Ann 10
Ochoa, Severo 188
Olot 132
Ombrédanne, Louis 191–3, 194, 225
On Monsters and Prodigies (Paré) 8
Opera Medicinalia (Bravo de Sobremonte) 17, 47
Opus Dei 198
Orfila, Mateo 78, 86–8, 89, 99
Organisation Internationale d'Intersexes (OII) 233
Ortega y Gasset, José 148
Otero Sánchez, A. 191–3, 194
Oudshoorn, Nelly 4, 10, 11, 19, 80, 224
Ovid 1, 38, 47, 53, 58

Palencia 205
Paracelsus 62
Parache, Dr 132
Pardo, Dr 182
Paré, Ambroise 2, 8, 47
Paris 89, 128, 139
Park, Katharine 6, 7, 8, 17, 32
Parsons, James 237
Partidas (Alfonso X) 56
Pascual, Salvador 141, 142–3, 145, 226, 230, 232
Paula de Ávila, St 58–9
Peiró, P. M. 86
Pérez, Juan Pablo 97
Pérez Cuadra, Dr 122
Perfecta casada, La (Luis de León) 84
Perpetua, St 61
Peset Cervera, Dr 102–3
Phlegon 38
Pichevin, R. 112–13
Piulachs, Pedro 193–4
Pla Messeguer, Teresa (La Pastora) 201–2, 226
Planelles, Dr 182

Plato 5, 47, 58
Pliny 37, 38, 53
Pons Tortella, E. 186–7
Pontano, Joviano 38
Pouchet, Gabriel 99–101, 102
Pozzi, Samuel 15, 78, 95, 103, 107, 109, 112, 114, 125, 127, 128–9, 139, 141–2, 144, 145, 179, 184, 185–6, 187, 191, 192, 194, 195, 197, 225, 226, 231, 238
Prevosti, Antoni 207
Prodigiorum ac Ostentorum Chronicon (Lycosthenes) 54
Puchol, Dr 122
Puebla de Guzmán 97
Pulido Martín, Dr 131

Quiero vivir mi vida (De Burgos) 150
Quintilian 14

Ravenna 54–5
Reifenstein syndrome 236
Reifferscheid, Dr 158
Revista Clínica Española 195
Revista de Medicina y Cirugía Práctcas 101, 103, 128
Revista de Occidente 149
Reyes Carrasco y Huelva, María de los 97–8, 107, 226
Riolan, Jean 8, 45, 48, 50, 237
Rivilla Bonet, José 49–51
Robert y Yarzábal, Bartolomé 109–10
Roda, E. 194–7
Rodrigo, J. 86
Rodríguez Lafora, Gonzalo 188
Rodríguez Zapatero, José Luis 233
Romanticism 84
Royo, Dr 131–2
Rubin, Gayle 10
Rueff, Jacob 54
Ruzicka, Leopold 190

Saint-Hilaire, Isidore Geoffroy 15, 78, 99, 100, 101, 106, 110, 238
Saldaña, Quintiliano 147
Salen, Dr 158
Salmacis 1, 5

Sánchez, Tomás 56
Sánchez Lucas, G. 163–5
Sánchez-Monge, Enrique 207
Sánchez y Sánchez, Manuel 101
Schiebinger, Londa 11, 111
Sebileau, Pierre 112, 113
Serrallach, D. N. 152
Settier, Alejandro 103–4, 105
Seville 122, 235
Sex, Gender and Society (Oakley) 10
Sex and Character (Weininger) 9, 10
Sexual Inversion (Ellis) 9
Siglo Médico, El 103, 129, 139, 155
Simarro Puig, J. 191–3, 194
Simpson, James Young 15, 106, 109, 112
Starling, Ernest H. 152
Steinach, Eugen 145, 151, 155, 182, 186
Stolberg, Michael 38, 45, 52, 53
Sumario de las maravillas, El ... (Gutiérrez de Torres) 54
Swyer's syndrome 236

Tardieu, Ambroise 89, 92, 93, 101
Teruel 186
Thomas, apocryphal book of 61, 62
Thompson, J. Arthur 149
Toledo 2
Torre Blanco, J. 161–2
Traité de gynécologie (Pozzi) 112, 127
Tratado de cirugía clinica y operatoria (Sebileau and Pichevin) 112–13
Tratado de medicina legal (Legrand du Saulle, Berryer and Pouchet) 101, 102
Tratado de medicina legal (Orfila) 86–8
Tratado de medicina y cirugía legal (Mata) 89, 90–4, 95–6, 103, 110
Turner's syndrome 123, 208, 236

Úbeda 40, 226
Ulibarri, Dr 91, 93–4, 95, 96, 226
Ulpian 56
Usandizaga, M. 163–5

Vade Mecum de medicina y cirugía legal (Mata) 89–90, 103
Valencia 103, 105
Valladolid 47
Valle Aldabalde, Dr 144
Valverde de Amusco, Juan 42
Vandebosch, Dagmar 146, 148
Varchi, Benedetto 237
Variation of Animals and Plants under Domestication, The (Darwin) 9
Varolio, Constantino 48
Venturi, Dr 158
Verh. der Deutschen Gesellschaft für Gyn. (Neugebauer) 112
Vesalio, Andreas 42
Vienna 109, 128, 139
Villabona 201
Vital Aza, I. 182
Vives, Juan Luis 84
Voronoff, Serge 151
Voz de Cádiz, La 235

Weber, Max 32
Weininger, Otto 9, 10
Weiss, Meira 234
Wilgefortis (Uncumber), St 58
Wilhelm, E. 140
Wittgenstein, Ludwig 31
Worbe, M. 93

Yáñez, Teodoro 99
Young, Iris 12–13

Zamora 37
Zawadowski, Dr 158
Zondek, Bernhard 155, 179, 182, 183, 184, 185